THE VOICEXML HANDBOOK
Understanding and Building the Phone-Enabled Web

Bob Edgar

Dialogic Corporation,
an Intel Company.

www.dialogic.com/voicexml

Published by:
  CMP Books, An Imprint of CMP Media Inc.
  12 West 21 Street
  New York, NY 10010

ISBN 1-57820-084-9

For individual orders, and for information on special discounts for
quantity orders, please contact:
  CMP Books
  6600 Silacci Way, Gilroy, CA 95020
  Tel: 800-500-6875, Fax: 408-848-5784
  Email: cmp@rushorder.com

Distributed to the book trade in the U.S. and Canada by:
  Publishers Group West
  1700 Fourth St., Berkeley, CA 94710

Manufactured in the United States of America.

Table of Contents

Preface

In the summer of 2000, I came across the VoiceXML 1.0 standard published by the VoiceXML Forum. I downloaded the specification and began to read it. I had been working on software development in computer telephony for more than 10 years, but I was completely baffled; I couldn't understand most of the specification. I had no idea what the motivation or underlying philosophy was — in short, I didn't "get" the language at all

Gradually, I realized my problem. I was the wrong kind of person. In particular, I wasn't a Web person — I had little experience of Internet technologies. The language was designed and described by people who lived and breathed Web servers and Web browsers, they took this background knowledge and conceptual framework for granted. So I decided to learn about TCP/IP, URLs, HTTP, HTML and graphical browsers. That made all the difference — now VoiceXML started to make sense. As I talked to people in the industry, I discovered that I wasn't alone, there were others who were interested but had trouble understanding VoiceXML: they didn't know enough about the Web, they didn't know enough about computer telephony, or they just wanted some other place to go if they didn't understand something in the standard.

That's when I decided to write this book. My first goal was to explain the concepts and technologies behind the Word Wide Web, telecommunications and computer telephony that you'll need to fully understand VoiceXML. Then I wanted to introduce VoiceXML step by step, starting with a simple "Hello, world" sample and gradually introducing new constructs in the language. Finally I wanted to present a comprehensive reference to VoiceXML in a complementary format to the engineering specification. By explaining the same things in a completely different way, I hoped to provide an alternative source of reference material for implementers and users of the VoiceXML language.

One critical decision to be made in writing this book was which version of the language to cover. An industry group called the

VoiceXML Forum published its VoiceXML 1.0 standard in March 1999. Subsequently the VoiceXML Forum asked the World Wide Web Consortium's (W3C's) Voice Browser Working Group to take over language evolution while the Forum concentrated on conformance and educational activities. On December 14th 2000, the W3C published its sixth draft of VoiceXML. As this book goes to press at the end of January 2001, that was the most recent draft available, and that is the version of the language described in this book. The draft describes its own status as follows.

This document is an unpublished Working Draft of the World Wide Web Consortium, and has been written for review by the Voice Browser working group. This document has been produced as part of the W3C Voice Browser Activity, but should not be taken as evidence of consensus in the Voice Browser Working Group. The goals of the Voice Browser Working Group are discussed in the Voice Browser Working Group charter.

Publication as a Working Draft does not imply endorsement by the W3C membership. This is a draft document and may be updated, replaced or obsoleted by other documents at any time. It is inappropriate to cite W3C Drafts as other than "work in progress".

This document is for public review.

Readers should be aware that the W3C may modify the standard, perhaps extensively, before its final acceptance and publication, and may not in fact publish any standard at all.

Readers should also note that there may be intellectual property claims inherent in the language or in applications created using the language. I recommend that you seek legal advice before implementing a browser or developing a VoiceXML application. You should not assume that you are free to use a language element or programming technique described in this book without an appropriate license.

Since the language specification was a work in progress, there were no VoiceXML 2.0 browsers available for me to use in developing and testing example code, so even if the language were finalized in its current form, there are probably errors and misunderstandings in the descriptions and samples presented in this book. I'll be posting errata and updates as the language evolves at:

```
http://www.dialogic.com/voicexml
```

Please visit this site to find out how to send corrections, suggestions for future editions of this book and other feedback, I'll be glad to review anything you send me.

I'm grateful to many people who spent their valuable time helping me with their patient explanations, discussions and feedback on early drafts of the book. In particular, I'd like to thank Jim Ferrans, Ted Glenwright, Dave Ladd, Jim Larson, Jeff Peck, Carl Strath-meyer and Moshe Yudkowsky. Any remaining errors and omissions are of course my responsibility.

Bob Edgar
Dialogic Corporation, an Intel Company
Sausalito, California
January 2001
http://www.dialogic.com/voicexml

1 The Telephony-Enabled Web

1.1   Calling the Web

This book is about VoiceXML, a new language which bridges two
great global networks: the public telephone network and the
World Wide Web. VoiceXML browsers provide an interface
between a caller on a standard telephone and an application
running on the Web.

With VoiceXML, traditional Web sites can build on their existing
infrastructure to add telephone access to their content and services
such as e-commerce purchases. With a Web-friendly technology
such as VoiceXML, it is a relatively small effort to provide a
telephone interface as an alternative to a graphical interface
because the layers underneath remain the same. In addition,
traditional automated telephone services (voice mail, banking by
phone...), which have historically been created from scratch as
separate applications, can be re-designed as a telephone interface
to a Web site. This can greatly reduce the amount of custom
development, special skills and parallel infrastructure needed to
offer services to telephone users and graphical browser users.

Imagine that you're in your car and new music you like is playing
on the radio. You use your carphone to call a voice portal, a
telephony Web site. A personalized portal recognizes that the call
is coming from your car, offers you driving directions, speaks the
latest news relevant to your business, updates you on your stock
portfolio and gives you a shopping link tied to your favorite radio
station so that you can order that music CD—all in one phone call.

VoiceXML is not the only way to access the Web through a phone.
Traditional languages such as Java can also be used. WAP
(Wireless Application Protocol) phones provide a small text
window which can be used for Web browsing using a language
called WML (Wireless Markup Language), however this requires
a special phone. Other languages similar to VoiceXML have been
proposed, however VoiceXML is so far the clear the clear leader.

1.2   The Voice Browser

Traditional automated call processing systems are developed using programming languages such as C++, Java and Visual Basic. While it is certainly possible to build telephony-enabled Web sites in this way, the telephony function calls (APIs) provided for such programs are rarely designed with the Internet in mind and the programming techniques used are often unfamiliar to Web developers.

The key to an Internet-friendly telephony architecture is the voice browser. A voice browser is a computer which answers a telephone call and interacts with the user. The browser outputs spoken prompts, which may be pre-recorded, streamed from a live audio source or generated using text-to-speech technology. Input is solicited from the user in the form of speech (which the browser processes using voice recognition technology) or touch-tone digits.

You can think of the telephone as the input/output device for the voice browser, playing the same roles as the screen, keyboard and mouse for a graphical browser.

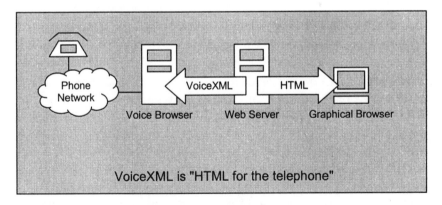

VoiceXML is a language for voice browser scripts. In short, it is an HTML for voice. VoiceXML is designed to be familiar to HTML developers and to exploit the Web infrastructure which is already in place to support graphical browsers.

The following table compares graphical and voice browsers.

Feature	Graphical Browser	Voice Browser
Language	HTML	VoiceXML
Browser output	Text and images laid out according to mark-up tags.	Streaming audio and synthesized text spoken according to mark-up tags, also pre-recorded sound files.
User input	Keyboard and mouse.	Touch-tone digits and spoken voice.
Resources retrieved from Web server	HTML pages, images, Java applets, ActiveX objects etc.	VoiceXML pages, speech recognition grammars, sound files, streaming audio.
Hyperlinking	Click on "hotspot" text or image, or submit form.	Say "hotspot" word (e.g. "help"), or submit form.

1.3   VoiceXML highlights

In this chapter we'll give you a quick tour of VoiceXML. We'll assume that you're reasonably familiar with Web and telephony concepts, but if you're not, don't worry: in later chapters we'll review this background material before we return to explore VoiceXML in full. (Note that the language described here is based on the December 14th, 2000 W3C draft of the proposed VoiceXML 2.0 standard, see the Preface for more information).

VoiceXML represents a radical departure from traditional telephony programming APIs using procedural languages like C++ and Java. VoiceXML is often descried as a dialog markup language, meaning that it instructs the computer on how to interact with a user through a conversation (dialog).

The key construct in VoiceXML is the voice form, also called a voice dialog. A form has a set of input fields, for example credit card number and expiration date. A VoiceXML form can be as simple as a list of fields; there may be no procedural code at all.

To interpret a form, the VoiceXML browser speaks prompts associated with fields in the form, collects user input to fill in values for each field, and submits those values to the Web server. In response, the Web server sends the next page. When the browser receives a new page, it starts presenting the first form on that page. A simple voice menu ("for this, press 1, for that, press 2...") is modeled as a form with a single input field.

Web developers will recognize a VoiceXML as being similar to an HTML form. When a VoiceXML form is completed (i.e., when all fields have been filled in), the values are sent to the Web server using the same HTTP feature used by a Submit button.

Visual Basic developers will also recognize the form metaphor. You can draw a VB form with input fields such a edit fields, check boxes and radio buttons, this form can be displayed and user input collected without adding any procedural code. The form is stored using a static, non-procedural definition (the *.frm* file). The Visual Basic run-time engine (analogous to the voice browser) has extensive built-in logic for handling user input (mouse clicks and keystrokes analogous to touch-tone and spoken words) in order to "fill in the form", i.e. select radio buttons and check boxes, enter text into edit fields and so on.

Database developers will recognize an analogy with SQL, which also replaced traditional, procedural APIs with a higher-level, non-procedural language.

1.4   The Web server as application server

When an automated telephony system is based on a voice browser, most of the application logic resides on the Web server.

For simple applications which don't need access to external APIs such as database access, it may be possible to write the VoiceXML pages in advance and store them on the server in the same way as static HTML pages. For example, audiotext applications might be done this way. (These are systems with touch-tone menu trees leading to pre-recorded information which rarely changes).

However, most applications will be based on dynamic content where at least some VoiceXML pages are generated by programs on the Web server. Such server programs use scripting technology such as ASP, Java Server Pages or the CGI Perl language.

This architecture allows the Web application developer to cleanly separate "business logic" (getting driving directions, buying a music CD...) from the "presentation layer" (page layout designed for a graphical browser user or spoken information designed for a voice browser user). The server logic for buying a book will be the same whether the credit card number and expiration date came from a graphical browser or voice browser. The module which executes the purchase transaction can report a simple success / failure result which is passed on to separate modules for conversion into HTML or VoiceXML, depending on the browser type. If this result is itself presented using an XML, then XML-oriented transformation technology such as XSL (XML Stylesheet Language) can be used to generate the HTML or VoiceXML as appropriate for the target browser.

1.5   An example form

Here is a simple VoiceXML form which collects a credit card number and expiration date from the user and submits those values to the Web server. The Web server will respond with the next VoiceXML page, which would presumably report the result of the transaction (say, by speaking "your card was accepted") then move on to the next form.

```
<form>
   <-- Input field for credit card number -->
   <field name="ccnr" type="digits">
      <prompt>
         Credit card number?
      </prompt>
   </field>

   <-- Input field for expiration date -->
   <field name="exp" type="date">
      <prompt>
         Expiration date?
      </prompt>
   </field>

   <-- "Submit button" -->
   </block>
      <submit next="getcc.jsp"/>
   </block>
</form>
```

A conversation with a VoiceXML browser interpreting this form might go like this.

BROWSER: Credit card number?

CALLER *(dialing)*:   012345678901234

BROWSER: Expiration date?

CALLER *(speaking)*: June two thousand three.

The form has three items: two input fields and one *<block>*, which is a special item used to contain executable code. In this case, the code in the block is a *<submit>* instruction.

The *ccnr* field has type *digits*, which is built into VoiceXML. The browser will to allow the user to dial touch-tone digits or speak the digits as "five four...". A well-designed browser will understand the various different ways people speak numbers: "oh" for "zero", "double six two five thousand", and so on, the details here will be browser-dependent.

The *exp* field similarly has the built-in type *date*, which expects the user to say or dial a date.

The form processes a field by first playing the field's prompt and then waiting for user input to specify the field's value. The fields are processed in the order they are written in the form.

The *prompt* tag speaks text using synthesized speech (text-to-speech).

To play a pre-recorded file rather than text-to-speech (which can sound rather robotic), use the *audio* tag and give the URL of the file in the *src* attribute. Our sample form re-designed to use pre-recorded prompts would look like the following.

```
<form>
    <field name="ccnr" type="digits">
        <audio src="ccnr.wav"/>
    </field>

    <field name="exp" type="date">
        <audio src="exp.wav/">
    </field>

    </block>
        <submit next="getcc.jsp"/>
    </block>
</form>
```

1.6   Speech input

User input is provided via speech and touch-tone digits. Speech input is processed using voice recognition technology.

In VoiceXML, acceptable speech input to a form, field or hyperlink is specified using a speech grammar. A speech grammar provides a template describing valid user input. A typical speech grammar specification looks something like BNF (Backus Naur Form), which is familiar to many programmers as a way to specify the grammar of a programming language. Built-in types such as *digits* and *date* have their own built-in grammars which are browser-specific (not defined by the VoiceXML standard).

Today's recognizers allow you to specify grammars written in simple text files using standard spelling for a spoken language such as English. Grammar compilers use spelling rules and phonetic databases to convert the (highly irregular, non-phonetic) spelling in English into phonetic spelling based on units called phonemes, which roughly correspond to individual spoken sounds as represented by letters (like "d" and "a"), or letter pairs ("th", "ee"). Compilation is fast enough that grammars can be updated in real time, or close to it.

VoiceXML 1.0 does not define a standard for speech grammars. Examples in the VoiceXML 1.0 specification are based on JSGF, Java Speech Grammar Format, but VoiceXML 1.0 browsers are neither recommended nor required to support this format.

VoiceXML 2.0 introduced SRGF, the Speech Recognition Grammar Format, for defining speech grammars. SRGF comes in two flavors: one based on XML, one based on ABNF (Augmented Backus-Naur Form). Support for the XML flavor of SRGF is required for conforming browsers, support for the ABNF form is optional.

Following is a simple grammar to get a credit-card type. This uses the optional ABNF form of SRGF, which is easier to read.

```
<grammar>
   visa
   | master [card]
   | amex
   | american [express]
</grammar>
```

The vertical bar " | " means "or", square brackets "[...]" represent optional word(s). This grammar will recognize any of the following utterances:

"Visa"
"Amex"
"Master"
"Master Card"
"American"
"American Express"

A form field using this grammar might look like the following.

```
<field name="cardtype">
   <prompt>
      Please say your card type.
   </prompt>
   <grammar>
      visa | master [card] | amex |
         american [express]
   </grammar>
</field>
```

1.7 Mixed-initiative forms

In the example form we have shown so far, the browser proceeds through the fields in the order they are specified in the page, playing prompts and getting user input one field at a time. This is called a machine-directed dialog since the computer determines the order in which values are collected.

VoiceXML also supports mixed-initiative dialogs, which allow the user to speak a single phrase which fills in one or more of the form's fields at one time.

For example, imagine a form which has three fields

- credit card type,
- credit card number,
- expiration date.

Suppose the user speaks:

"My credit card number is 1234567890, expiring October 2004."

The browser recognizes that two fields are filled in, then proceeds to prompt and get any unfilled fields, as for a machine-directed form. In this case, the credit card type has not been provided, so the browser would proceed with the credit card type prompt, such as "Please say the type of your credit card".

To create a mixed-initiative form, a grammar is specified for the form (as opposed to a grammar for a field).

1.8   Touch-tone input

VoiceXML also supports input using touch-tone digits. Grammars written for touch-tones are very similar to those using spoken words. As you would expect, you represent touch-tones by writing digits, "#" or "*" where you would write a word in a grammar for matching speech.

1.9   Recording speech input

The *record* tag instructs the browser to record audio input.

```
<record name="Message0433456" beep="true"
   maxtime="240s" finalsilence="3000ms"
   dtmfterm="true" type="audio/wav"/>
```

The audio data for the recording is stored locally, most likely, on the browser's hard drive. The audio can be played back or submitted to the Web server for permanent storage.

1.10   Speech output

There are three major categories of speech output: pre-recorded, streaming and synthesized.

Pre-recorded speech uses sound files which are familiar from desktop formats such as Wave. Sound files are specified using URLs in an *<audio>* tag, for example:

```
<audio src="Hello.wav"/>
```

Streaming audio (i.e., audio provided in real time by the Web server) is requested by disabling caching, forcing the browser to fetch the audio data each time the URL is referenced:

```
<audio src="LiveNews.wav" caching="safe"
   fetchhint="stream"/>
```

Synthesized speech is often called Text-to-Speech or TTS. TTS technology is improving, but in most cases still produces a somewhat robotic voice.

VoiceXML 2.0 introduced a standard mark-up for text known as the Speech Synthesis Markup Language, SSML. The developer may elect to use SSML instructions to specify speed, volume, pitch and emphasis of the verbal rendering of text. Or the application can simply provide text and let the speech synthesis engine use default values for these and other parameters.

Here is a text-to-speech prompt which speaks a variable value and uses the SSML *<emphasis>* tag which emphasizes (stresses) a segment of the speech. You can think of this as similar to bolding text in a graphical browser.

```
<prompt>
   <emphasis>Please</emphasis> buy my product.
</prompt>
```

1.11   Hyperlinks

Moving to a new dialog or page is called hyperlinking, and is similar to clicking on a "hotspot" (e.g. underlined text) on an HTML page.

Each dialog must link to a new dialog at some point; otherwise the browser session ends. Typically this means that the browser will hang up on the user.

A hyperlink can be specified using a *<submit>* tag, which works like a Submit button in an HTML form, or a *<goto>* tag, which specifies the target URL without sending any variable values to the server.

The hyperlink can specify a new dialog (form or menu) on the same page (by using the #*name* syntax familiar from HTML), a new page, or a new dialog on a new page. If a new page is specified without naming a dialog, the browser starts processing the first dialog on the page.

Hyperlinks can also be specified using the <*link*> tag, which specifies a grammar that remains active while the browser processes a dialog. This feature can function like a navigation bar on a graphical Web page: for example, it could allow the user to say "stocks" at any time to transition to a stock quotes page.

Here is a link tag which responds to the touch-tone digit star "*" or the spoken words "main menu".

```
<link next="MainMenu.vxml">
   <grammar>
      main menu
   </grammar>
   <grammar mode="dtmf">
      "*"
   </grammar>
</link>
```

1.12   Root document

VoiceXML applications can define a root document. This is a VoiceXML page which remains in scope while other pages are swapped out due to hyperlinking. A typical use of the root document is to define grammars which respond to user input, thereby providing features similar to a navigation bar in a graphical site. Such grammars can be used to respond to words such as "Help" or to transfer to other areas of the site ("E-mail", "Shopping").

1.13   Client-side scripting

VoiceXML includes simple decision-making features including <*if*>, <*else*> and <*goto*> tags and, for more complex tasks, a full implementation of ECMAScript (sometimes called JavaScript or JScript), a powerful procedural language. This allows VoiceXML pages to make sophisticated decisions without having to involve the Web server.

By design, these scripting features are for user interface customization only, not for adding additional APIs such as database access.

1.14   VoiceXML minimizes client / server interaction

In traditional telephony programming, each low-level user input action (touch-tone digit, spoken word) must be reported to the application server. In VoiceXML, the client voice browser processes most user input without involving the application server. The server sees only the field values submitted by the browser.

A single VoiceXML page may contain multiple dialogs. By including several dialogs per page, application developers can further reduce the client/server round-trips, minimizing the load on the Web server and improving responsiveness.

1.15   VoiceXML is standards-based

VoiceXML is an XML markup language. Programmers familiar with XML or HTML find the style of VoiceXML familiar.

VoiceXML is based on features already available in existing Web protocols, especially HTTP. Resources such as VoiceXML pages, pre-recorded sound files and voice recognition grammars are stored on Web servers in exactly the same way as GIF and JPEG images for graphical Web sites. Existing Web servers will support VoiceXML without difficulty.

A wide range of XML development and deployment tools is available: syntax checkers, code generators, debuggers, interpreters, accelerators and more. These tools can assist in developing VoiceXML browsers and in creating VoiceXML infrastructure such as "server farms".

1.16 VoiceXML portability

VoiceXML 1.0 provides some portability across browsers, however some significant features are browser-specific. In particular, the syntax of voice recognition grammars is implementation-defined in VoiceXML 1.0.

VoiceXML 2.0 is designed to address these issues and should provide a high degree of portability among compliant browsers.

1.17 Extensions to VoiceXML

VoiceXML allows browser vendors to provide language extensions through the <object> tag (not by adding new tags to the language!). The syntax of this tag is defined by the language, but not the interpretation of the attributes, this is up to the browser vendor to define.

1.18 Desktop voice browsers

Voice browsers are not limited to serving telephone callers. A voice browser might also be found on a desktop PC to enable Web-browsing for sight-impaired users. However, it is the business potential of providing enhanced telephone services that has the industry excited today and that will be the main focus of this book.

1.19 What's missing from VoiceXML 2.0?

VoiceXML is a new language which is only now being deployed widely. Time will show which additional features will be demanded and added to future revisions of the language, but we can already make an educated guess at some of the areas where future extensions may be defined.

1.19.1 Telephony Control

You may have be wondering when we would introduce tags like <waitforcall> and <answercall>. There are no such tags. It is entirely up to the VoiceXML browser how a user session gets started. HTML does not specify that you should start your browser by double-clicking a desktop icon, VoiceXML does not specify how or when a telephone call starts interacting with the browser, nor what should happen when the browser session ends. There is a *<disconnect>* tag to hang up the call and a *<transfer>* tag, which instructs the browser to transfer the call; this is the only telephony control currently in VoiceXML 2.0. Some VoiceXML users have requested enhancements to enable more telephony features such as conference calling; efforts are under way at the time of writing (January 2001) to draft proposals for extensions to VoiceXML along those lines.

1.19.2   Semantics

In voice dialogs, the term semantics refers to extracting data fields (called slots) from user input. For example, suppose the user speaks the following phrase.

```
"I want to buy a ticket to fly from Boston to New
York on December fifteenth."
```

The semantic content of this phrase might be expressed as the following *slot=value* pairs:

```
Command=BuyTicket
FromCity=Boston
ToCity=New York
Date=12/15
```

The process of extracting semantic content from user input can be achieved by providing mark-up for grammars. The VoiceXML 2.0 draft specification available at the time of writing leaves significant gaps in this area, which would require each browser vendor to define proprietary solutions; however the editors of the standard are hopeful that this issue can be addressed.

1.19.3   N-best recognition results

Most voice recognition technology can provide more than one candidate match to a given word or phrase. For example, the recognizer might report two possible matches for a word: "Fee" with 70% probability, "See" with 30% probability. VoiceXML 2.0 currently does not provide a standard way to return alternative results to the application.

1.19.4   Specialized Speech Recognition

Speaker identification (identifying a speaker by recognizing his or her particular voice) is not supported, nor is speaker verification (speaker identification which resists false-positives).

Standardized VoiceXML grammars do not support the fine granularity necessary for detecting speech-impaired input from disabled speakers or foreign language speakers that may be required for "learn to speak" applications.

1.19.5   Talking Heads

There are no mechanisms in SSML to synchronize a talking head with synthesized speech. This feature is more applicable to desktop than to telephony applications.

1.20   History of VoiceXML

Several companies developed markup languages for voice browsers in the mid to late 1990s. IBM's original language was SpeechML. AT&T Bell Labs developed PML (Phone Markup Language) which evolved into two different flavors used by Lucent and AT&T. Motorola created VoxML and HP Research Labs created TalkML.

IBM, AT&T, Lucent, and Motorola combined to found the VoiceXML Forum and created VoiceXML (briefly known as VXML). The preliminary specification was published as VoiceXML 0.9 in August 1999, with version 1.0 published in March 2000. The VoiceXML Forum asked the World Wide Web Consortium's (W3C's) Voice Browser Working Group to take over language evolution while the Forum concentrated on conformance and educational activities. In January 2001 the W3C published drafts of VoiceXML 2.0, the Speech Recognition Grammar Format, and the Speech Synthesis Markup Language.

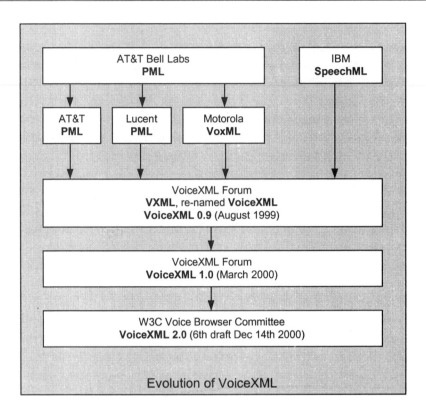

Evolution of VoiceXML

2 Web Browsing

2.1   How a Web browser works

Suppose you've just typed a URL such as *http://www.intel.com-/index.htm* into your browser. What happens next?

First your browser has to find the IP address of *www.intel.com*. This is done using an Internet protocol called Domain Name Service (DNS). The details aren't important to us; it's enough to know that DNS reports Intel's Web server address to be 192.102.197.190.

Next, the browser establishes a TCP/IP session with the server at address 192.102.197.190. TCP/IP is a protocol that allows two computers on a network to send a stream of bytes to each other without worrying about network data errors. To establish a session you need both an IP address and a port number. The port number is between 1 and 65,535 (FFFF hex) which identifies the type of software that you want to talk to. In our case the port number will be 80, which is the default port for HTTP (we'll explain more about what HTTP is shortly).

Once the session is established, the browser asks the server for the data which is named by the URL. A URL is composed of three main parts: a protocol name (*http://*), a server name (*www.intel.-com*) and a file name (*/index.htm*). The browser asks for the data by sending a *GET* command to the server requesting the contents of that file name. The command is a text string, which could be as simple as this:

```
GET /index.htm HTTP/1.0
```

Intel's server responds by sending the data in the file named */index.htm*. We're going to pretend that Intel has a very simple home page, so the server sends:

```
<HTML><BODY>Hello from <B>Intel</B>.</BODY></HTML>
```

The text is formatted using HTML, the HyperText Markup Language. The text to be displayed is "marked up" by tags such as

<B>, which displays text in bold until the matching end-tag *</B>* is found. The browser displays:

Hello from **Intel**.

GET is the most important command defined by HTTP, the HyperText Transfer Protocol. The idea is very simple: the client (browser) says "please send me the data in this file name", the server responds by sending the data. HTTP is a clever invention by one man (Tim Berners-Lee) — it is not at all obvious, except perhaps in retrospect, that such a simple protocol, plus a simple layout language like HTML, can drive a global hypertext database. (HTML borrows from an earlier document language called SMGL; HTTP is Tim's own). Other commands include *HEAD* (requests information about a file), and *POST* (sends data from the client to the server). There are a few other commands, but they are rarely used.

2.2   HTTP command syntax

HTTP 1.0 supports a minimal *GET* command which just specifies the file name and protocol version:

```
GET filename HTTP/1.0
```

Here is a more typical *GET* command using HTTP 1.1, which is currently (January 2001) the most widely used version of HTTP:

```
GET /index.htm HTTP/1.1
Accept: text/html, */*
Accept-Language: en-US
User-Agent: Mozilla/4.0 (compatible; MSIE 5.01;
  Windows NT)
Host: www.intel.com
Connection: Keep-Alive
```

An HTTP command is, in HTTP jargon, called a client request. A client request consists of the following parts, some of which are optional.

2.2.1   Method

This is what we've been calling a "command". The syntax of a method is always:

```
methodname filename HTTPversion.
```

2.2.2   Request-Header

This tells the server more about the browser client, including optional fields such as *User-Agent*, which gives information about the browser version, and *Accept*, which tells the server what data formats the browser understands. A required field (in HTTP 1.1) is *Host*, which identifies the URL originally used to find the server. This is useful for servers which host multiple Web sites; the *Host* field identifies the site and therefore which sub-tree of the server's logical file system to use when processing the request.

2.2.3   Entity-Header and Entity-Body

If the command sends data, then the Entity-Header and Entity-Body sections must be included. The Entity-Body is the data itself, as far as HTTP is concerned it is just an array of bytes which is not to be analyzed further. The Entity-Header has two fields: *Content-Type*, which gives the Media Type (MIME type) of the data to follow, and *Content-Length*, which gives the length of the data in bytes. A typical use for an Entity-Body is to send the values of form variables as the result of a Submit button. Here is an example:

```
Content-Type: application/x-www-form-urlencoded
Content-Length: 27
CardNr=0123456789&Type=Amex
```

In HTTP 1.1, a *GET* command must include the method and a *Host* Request-header; the other fields are optional.

2.2.4   The Accept Field

The *Accept* field of the Request-Header lists the file formats that the browser understands, in order of preference. Consider this:

```
Accept: image/gif, image/jpeg
```

The browser is saying: "I'm ready to accept a GIF or a JPEG file in response to this request, I'd prefer to get a GIF if possible." In HTTP terminology, a file format is a Media Type. A Media Type is written in one of these forms:

```
type/subtype
type/*
*/*
```

As you would expect, the star (*) is a wild card which means anything accepted, so */* means "accept any file format". You will sometimes see a Media Type referred to as a MIME Type. MIME (Multipurpose Internet Mail Extensions) is a standard which describes how e-mail messages and attachments are formatted. For our purposes, Media Types and MIME Types are the same thing.

2.3   HTTP server response

The syntax of a server response is similar to that of a client request. Here, now including the header fields, is a full response to a *GET* request:

```
HTTP/1.0 200 OK
Server: Apache/1.1.2
Last-Modified: Sun, Dec 24th 2000 22:13:02 GMT
Connection: close
Content-Type: text/html
Content-Length: 50
<HTML><BODY>Hello from <B>Intel</B>.</BODY></HTML>
```

The first line is:

```
HTTPversion statuscode reason
```

The status code reports the success or failure of the command, the range 200 - 299 always indicates success. The *reason* phrase is a short string explaining the code. Error code 404 will no doubt be all too familiar, it is reported by the server like this:

```
HTTP/1.0 404 Page not found
```

The Entity-Header section describes the data, which in this case reports that the data to follow is 50 bytes of HTML text. The Entity-Body section comes last and contains the data itself.

2.4   HTTP GET, POST and HEAD commands

With these preliminaries out of the way, we can now look in more detail at the principal HTTP commands: *GET, POST* and *HEAD*.

GET requests file data. It has no Entity-Header or Entity-Body; it does not send any information except the header fields.

POST also requests file data, it is like *GET* plus it sends an Entity-Header and Entity-Body to the server. As we shall see, this can be used for sending data from a form (a feature exploited by VoiceXML for sending user input) or for uploading an entire file (used by VoiceXML for voice recording).

HEAD is exactly like *GET*, except that the server responds only with header information, the file data is not sent. This is useful for checking fields such as Last-Modified, which tells the browser whether it is safe to use data stored in its cache (faster) or whether it should request a fresh copy of the file data (slower but safer).

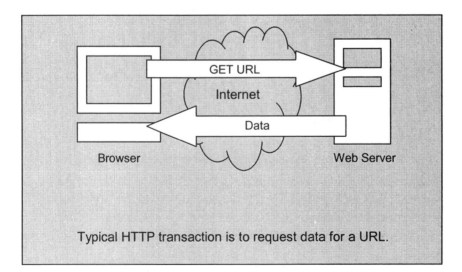

Typical HTTP transaction is to request data for a URL.

2.5 Static and active content

What does an HTTP server do when it receives a request for a file? Sometimes the server will search for a file on its hard drive and copy the data in that file to the browser. But that's not the only possibility. The server must translate a filename into data, how exactly that happens is entirely up to the server. Instead of copying a file, the server may run a program and send the output from that program to the browser. If the data is stored in a file, this is called static content. If the data is created by running a program, it is called dynamic content. Programs which output data to satisfy HTTP requests are often called server scripts.

How does the server know whether to copy a file or run a script to generate the data? The details depend on the server, but very often the filename extension is used. If the filename ends in *.htm* or *.html*, the server assumes this is a static content file and that the Media Type is *text/html*. If the filename ends in *.pl*, it assumes the file contains a CGI (Common Gateway Interface) script written in the Perl language. Other common extensions include *.asp* for scripts written using Microsoft's Active Server Page language and *.jsp* for Java Server Pages.

The difference between static and active content is invisible to the browser: it gets an HTML page (or GIF image or whatever), it does not know or care whether that page was found in a file or was output by a script.

Let's see how this process works in one very widely used Web server: Apache. Suppose an Apache server receives the command:

```
GET /index.htm HTTP/1.0
```

The main configuration file for Apache is called *httpd.conf*, which is a text file where the server administrator can enter and edit parameters. Assuming we're running the Windows version of Apache, one important entry in this file might appear as follows:

```
DocumentRoot "d:/DocRoot"
```

When a file name is requested, Apache pre-pends the
DocumentRoot value to give the full path name, so in this example
a request for *index.htm* would translate to the Windows path name
d:/DocRoot/index.htm. Another important Apache configuration file
is called *mime.types.conf*. A typical entry in this file is:

```
text/html html htm
```

This says that files with extensions *.htm* and *.html* have the default
Media Type (a.k.a. MIME type) *text/html*.

By default, Apache doesn't use the filename extension to
recognize scripts; instead it uses a directory name. This is
specified by a parameter called *ScriptAlias* in *httpd.conf*, which
might look like this:

```
ScriptAlias /cgi-bin/ "d:/web/CGIScripts"
```

This says that if the filename starts with */cgi-bin/*, run the script
under *d:/web/CGIScripts*. Consider the following client request:

```
GET /cgi-bin/login.pl HTTP/1.0
```

In response, Apache would attempt to execute a script in
d:/web/CGIScripts/login.pl.

2.6   HTML, HyperText Markup Language

If you're reading this book, you probably understand at least the
basics of HTML already, but we'll give you a quick introduction
just in case.

HTML is a page layout language. A page of HTML is a series of
instructions to a graphical browser which specify how text and
images are to be displayed.

An HTTP server sending the pages knows almost nothing about
HTML, all the server has to do is locate a file or script given the
filename from a URL and know the Media Type (*text/html*, in this
case).

Let's return to our minimal example, our imaginary Intel home page:

```
<HTML><BODY>Hello from <B>Intel</B>.</BODY></HTML>
```

All HTML pages are enclosed within *<HTML>* and *</HTML>*. This pattern—a pair of tags enclosing some information—is widely used in HTML and is a model for XML, as we shall see later. Most pages will also include a *<HEAD>* section, which gives descriptive information such as the title, which is found between *<TITLE>* and *</TITLE>*:

```
<HTML>
    <HEAD>
        <TITLE>
            Intel's Home Page
        </TITLE>
    <HEAD>
    <BODY>
        Hello from <B>Intel</B>.
    </BODY>
</HTML>
```

We formatted the source code using indents to show the structure of the document more clearly. In HTML, any amount of white space (blanks, tabs and end-of-line characters) is usually considered to be equivalent to a single space, so we are free to lay out the page in any way which is convenient for a human reader. If you want to force a line break, you can use the *
* tag. The *<P>* tag ends a paragraph, which most browsers will display by ending the current line and inserting an additional empty line.

The *
* tag is an example of a tag which is not paired, there is no *</BR>*. The *<P>* tag is supposed to be paired—strictly speaking, you should enclose a paragraph inside *<P>...</P>*; however many authors and HTML editors don't bother with the *</P>* and most browsers allow this indiscretion.

The spirit of HTML is one of tolerance: a browser should try to interpret a page as best it can even if there are tags it doesn't recognize (which should be silently ignored) or HTML syntax errors (say, a missing ">"). XML, as we shall see, couldn't be more different in this regard — a zero tolerance policy is strictly enforced.

Tags are also used to request text formatting. For example, text found within <B>...</B> should be rendered in bold, <U>...</U> underlined, and so on. The size of the text can be controlled using the <FONT> tag, for example:

```
<FONT size="1">The fine print</FONT>
```

This introduces a new element of HTML: an attribute, in this case *size="1"*. Text sizes range from 1 (smallest) to 7 (largest), so this example requests the smallest size. Attributes appear within the angle brackets following the tag name, and always have the form *name=value*. Tag values may be enclosed in double quotes "..." or single quotes '...'. Many browsers, in keeping with HTML's spirit of tolerance, allow you to omit the quotes and just write something like *size=1*, though this is not strictly speaking legal. Multiple attributes can be specified separated by white space, for example:

```
<FONT size="7" color="#0000FF">Big Blue</FONT>
```

In HTML, tag names and attribute names are not case-sensitive, so <FONT>, <Font>, <fONt> and <font> are all acceptable. The above examples use upper-case tag names and lower-case attribute names, but this is purely a matter of taste and style.

2.7   Hyperlinks

A hyperlink is a text fragment or image which allows the user, usually via a mouse click, to jump to a new location. In most cases, the hyperlink causes the browser to fetch and display a new page. The underlined text or active image is called a hotspot. Most graphical browsers change the cursor to a distinctive shape (often a hand) to give visual feedback alerting the user that the cursor is on a hotspot.

In HTML, a hyperlink is defined by an <A> tag, for example:

```
<A href="http://www.intel.com/index.htm">
    Intel</A>
```

This specifies that the text *Intel* is a hyperlink to Intel's home page. If the hyperlink is text, as in this example, it is usually called out in some way, most often by underlining. (For historical reasons, the tag name *A* stands for "anchor").

The *href* attribute specifies the destination URL. There are two types of destination URL: global and relative.

The URL for a global hyperlink includes a server name, as in the above example.

A URL without a server name is called relative. For example,

```
<A href="index.htm">Home</A>
```

The browser must request a full path name from the server because HTTP does not support relative names. To construct the full path from a relative link, the browser keeps track of where it is in the server's path name tree. The document can override this by specifying an *href* attribute in a *BASE* tag, which must be placed in the document header (i.e., inside *<HEAD>...</HEAD>*).

An anchor is a label which identifies a position within an HTML page. It is sometimes called a fragment. An anchor is inserted by using a variant of the *A* tag which uses a *name* attribute, for example:

```
<A name="Top">This is the top of the page.</A>
```

The anchor name is appended to a URL at the end following a pound sign "#". Consider the following hyperlink:

```
<A href="#Top">Go to top</A>
```

Clicking on this link asks the browser to jump to the anchor *Top* within the current page. You can specify anchors in global or relative links. Here is an example of an anchor in a global link:

```
<A href="http://www.intel.com/index.htm#Bottom>
   Go to bottom of Intel's home page</A>
```

2.8   URL, Uniform Resource Locator

A URL, (Uniform Resource Locator), is the name of a resource (HTML page, image or other data) on an IP network. Technically, a URL is a special case of a more general concept called a URI (Uniform Resource Identifier), but the term URL is more familiar and is the one we will use in this book.

The syntax for URLs is familiar to anyone who has used a Web browser. The general form is:

[scheme:] [//servername] [:port] [filename] [#anchor]

All these parts may be optional or forbidden, depending on the context.

2.8.1   Scheme

Scheme is the correct technical term for what is usually called the protocol: HTTP, FTP (File Transfer Protocol) and so on.

2.8.2   Server Name

The *servername* part (sometimes call the *hostname*) is specified in one of two ways: as a dot-separated sequence of domain server names, in order from lowest to top-level domain, or using dotted decimal notation, which specifies a 32-bit IP address using four decimal (base-10) numbers in the range 0 to 255, one for each of the four bytes in a 32-bit value. For example, the *servername* for the Intel Web server can be specified as *202.153.118.46* or as *www.intel.com*.

If the name rather than the dotted decimal address is specified, then Domain Name Service (DNS) is used to translate the name to an IP address.

Dotted decimal addresses may be slightly faster (since no name lookup is required) and will work even if DNS is not available. However, the IP address of a server is more likely to change than its name.

If DNS is available, you can easily find the IP address of a name by using the command-line utility *nslookup*, which is found on most computer systems which are on the Internet. For example, try typing the following at a command prompt:

```
nslookup www.intel.com
```

The response should be a message like:

```
Name: www.intel.com
Address: 202.153.118.46
```

By convention, the server name *localhost* is reserved for the computer where the application is running. This is useful for development and testing where the client and server are running on the same PC. For example:

```
http://localhost/TestPortal.vxml
```

2.8.3   Port Number

The port is a number, specified as decimal digits (not octal or hex!), which is required to set up an IP session between the Internet server and its client. Each scheme has a default port number, so you usually don't need to provide one. For example, FTP defaults to port 21 and HTTP defaults to 80. These are called well-known port numbers. Examples of URLs which explicitly give the port number look like this:

```
http://www.intel.com:80
http://www.intel.com:80/com/index.htm
http://202.153.118.46:80/com/index.htm
```

2.8.4   File Name

The filename part specifies a particular file or script on the host server. As we have seen, if the filename is omitted, the root (top-level file, named "/") is assumed and, in the case of HTTP, the server will respond with some pre-defined file name as the home page (typically index.htm or index.html, but this is up to the Web server to figure out).

2.8.5   URLs for FTP

FTP is File Transfer Protocol, which is a simple protocol which allows a file to be sent between a client (e.g., a browser) and a server in either direction.

FTP allows a user name and password to be specified as part of the URL using the following syntax:

```
ftp://[UserName[:Password]]@hostname[path]
```

For example,

```
ftp://AUser:Secret@ftp.intel.com/Files/Readme.txt
```

The password may be omitted, for example:

```
ftp://AUser@ftp.intel.com/Files/Readme.txt
```

If the FTP user name is omitted, *anonymous* is used, in which case the password should be either the user's e-mail address or omitted.

2.8.6   URLs for Local Files

The *file:* scheme gives you direct access to a file system on the client. The server name defaults to *localhost*, the computer where the browser is running. For example, the following would specify a file on a Windows system:

```
file://C:\winnt\system.ini
```

On UNIX, a file URL might look like this:

```
file:///users/James.Roberts/MyFiles/Readme.txt
```

2.8.7   Anchors

An anchor is a named location within a file, as we saw in the earlier discussion of hyperlinks. Anchors are sometimes called fragments. Anchors are used by clients (e.g. browsers) but not by servers. The client will always strip out an anchor before sending a URL as part of a command.

2.9   Cookies

A cookie is a string of text which is sent from an HTTP server to a browser.

The browser stores the most recent cookie from each server. Each time the browser sends a command to a server, it checks its list of active cookies. If there is a cookie for that server, the browser includes it in the request.

The server creates a cookie by including a *Set-Cookie* field in any response, for example:

```
HTTP/1.0 200 OK
... other response header fields here...
Set-Cookie: name=James Roberts; domain=intel.com;
   expires=Sun, Dec 24th 2000 23:59:59 GMT
```

If the browser sends a request to *intel.com* any time before midnight on 12/24 then it should include this cookie. The request could look like the following:

```
GET /index.htm HTTP/1.0
... other request header fields here...
Cookie: name=James Roberts
```

Cookies are formatted like this:

```
property=value;property=value;...
```

The *domain* property indicates the domain name of the server.

As this example illustrates, cookies go bad — they have a use-by date. The expiration date of a cookie is given by its *expires* property. If no expiration date is given, the cookie is deleted when the browser closes.

Cookies are not part of the official HTTP standard, but they are supported by many browsers.

2.10   Session tracking

The concept of a session is important and also confusing, because there are at least three different kinds of session between a browser and an HTTP server: a TCP/IP session, a browser instance session, and a transaction session.

2.10.1   TCP/IP Session

A TCP/IP session is a connection between a client (browser) and a server which enables them to send streams of bytes to each other. Even if the user does not leave a given Web site, the browser may need to re-establish a new session in order to get the next page.

An HTTP request can ask the server to keep the TCP/IP session open by using this General-Header field:

```
Connection: Keep-Alive
```

If the server accepts, it will reply with the same field. However, the server is under no obligation to do this, and may respond with:

```
Connection: Close
```

Even if the server does accept, it may time-out after some pre-set interval in order to free up resources, requiring the client to start a new TCP/IP connection.

2.10.2 Browser Instance Session

An instance of a browser is a running copy of the browser software. A typical browser session on a PC is started by double clicking on the browser icon, or by clicking on a hyperlink in a document (an e-mail message, for example). A given user may have several copies of the browser open at once, each one is (by definition) a separate browser instance session.

2.10.3 Transaction Session

By a transaction session we mean a logically connected series of interactions between a user and a Web site. (The industry has not settled on a term for this concept, so don't assume other people are going to know what this means). For example, a transaction session with a shopping site could mean putting items into a shopping cart, entering a shipping address and credit card number, and finally confirming the purchase. If the user does not complete a transaction before the TCP/IP or browser instance session ends, the site would like to keep the items in the shopping cart and any information already entered, such as the shipping address, so that the user doesn't have to start from scratch when he or she returns to the site.

It is easy, but uninteresting, for the server to know when a TCP/IP session starts and stops. More useful, but harder to do with HTTP, is to recognize browser instance sessions or transaction sessions. It is also useful to recognize a given user when he or she returns to a site that they have visited before (in this case you could say that the entire relationship between the site and a customer is one long transaction session).

Cookies provide a simple and convenient solution for many session tracking needs. If you don't specify an expiration date, the cookie will live as long as a browser instance session. If you specify an expiration date far in the future, the cookie can track a transaction session. However, if you use cookies, you generally have to make one very important assumption: there is only one person using the browser. For a graphical browser, this is often a good assumption, though not always (e.g., the PC is used by the public at a local library). A VoiceXML browser is like a library PC: chances are that many different users will access it, so cookies are of limited usefulness.

If the browser has many users, does not support cookies or has cookies disabled as a user preferences, then there is a technique called URL embedding which allows tracking within a browser instance session. Suppose you have asked the user to log into your site and assigned or retrieved the account number, 1234. The idea is to embed 1234 into every hyperlink. The following simple example shows how this works.

The home page asks the user to log in.

```
            Store Home Page


     Name:   | James       |


  Password:  | Roberts     |
```

The server looks up the user name and replies with a new page which is customized for that user:

```
            Welcome, James.

          Click here to buy books

          Click here to buy music
```

All hyperlinks in the new page (in this example, the links behind *Click here to buy books* and *Click here to buy music*) include the user's

account number. For example, the URLs might be
/1234/Scripts/BuyBooks.jsp and /1234/Scripts/BuyMusic.jsp. If the
user clicks on the first link, the server will actually run
/Scripts/BuyBooks.jsp and pass 1234 as a parameter to the script.
The new page sent from *BuyBooks.jsp* will do the same thing, it
will pre-pend /1234 to all hyperlink URLs. While workable, this
technique is far from ideal: it requires customization of the server
software to manipulate URLs, and it means that *all* internal links
in *all* pages must be edited before sending (otherwise you lose
track of the session), which may have a significant performance
penalty.

2.11   User input and forms

Forms (also called dialogs) are a key construct in VoiceXML. We
now move on to explain how forms work in graphical browsers,
emphasizing the concepts relevant for VoiceXML.

Simple hypertext does not provide for user input: you can click on
a link, but you can't enter information. If you want to create a
Web-based application such as an invoicing system or search
engine which needs text input, what are your options? At the
opposite extreme from static text are solutions such as Java
applets and ActiveX controls — code which is downloaded to the
browser and executed on the client computer. This is powerful,
but complex. A happy medium, sufficient for many applications,
is to use a form. HTML supports simple forms, so you don't have
to learn a new language to create them and the browser doesn't
have to support a new language to execute them.

2.12   A login form

Let's consider a simple example: a login screen which prompts the
user for a name and password. Here is such a form displayed in a
browser.

The HTML source code for this page is as follows.

```
<HTML>
    <HEAD><TITLE>Login</TITLE></HEAD>
    <BODY>
        <FORM action="/cgi-bin/login.pl">
            <INPUT TYPE="text" name="user">
                User Name<BR>
            <INPUT TYPE="password" name="pwd">
                Password<BR>
            <INPUT TYPE="submit" name="Login">
        </FORM>
    </BODY>
</HTML>
```

A form is defined inside *<FORM...>...</FORM>*. This form has three items, each defined by an *<INPUT>* tag. There are several kinds of form item; in this example we see a standard text input field (specified by the *type="text"* attribute), a text input field where characters are echoed as stars (*type="password"*) and a submit button (*type="submit"*).

The submit button is special — every form must have one. When the submit button is clicked, the data from the form is sent to the server as part of an HTTP command.

By default, a submit button uses the HTTP *GET* command. The input fields are sent by appending their values to the URL in the following way. Suppose the user entered "James " into the username field, and "Secret" as the password. When the *Login* button is clicked, the browser sends a *GET* command with this URL:

```
/cgi-bin/login.pl?user=James&pwd=Secret
```

(As an aside, this shows that the password field is not very secure: anyone who intercepts the command for example by eavesdropping on your TCP/IP session can read your password).

The filename part of the URL, */cgi-bin/login.pl*, comes from the *action* attribute of the *<FORM>* tag:

```
<FORM action="/cgi-bin/login.pl">
```

The form data is appended to the filename using this syntax:

```
filename?name=value&name=value...
```

The name of the field comes from the *name* attribute, as in:

```
<INPUT TYPE="text" name="user">
```

2.13 URL encoding

Some special characters, including most importantly spaces, are not allowed in URLs. This presents a problem: suppose that "James Roberts" was entered as the user name, how can the browser send this?

The answer is a technique called URL encoding. Special characters are replaced by %*xx*, where *xx* is two hexadecimal digits giving the ASCII value of the character. Space is hex 20, so the URL would look like this:

```
/cgi-bin/login.pl?user=James%20Roberts&pwd=Secret
```

Some (mostly older) browsers use a plus sign "+" to encode a space, so you may find that the URL appears as:

```
/cgi-bin/login.pl?user=James+Roberts&pwd=Secret
```

Use of "+" is discouraged, browsers should use %20 instead, but if you are writing a server script to process form data then it is prudent to support both.

2.14   Server scripts

Forms require a server script to process the data. In our example, the script is *login.pl*, which suggests that it is written in the Perl language. Perl is a scripting language widely supported on Web servers.

When the server receives a URL with variables, the names and values are stored and made available to the script. How this is done depends on the scripting language and, in some cases, the server. In Perl, the values are passed in via the standard environment variable *QUERY_STRING*. In our example, it would be set as follows:

```
QUERY_STRING="user=James%20Roberts&pwd=Secret"
```

Notice that it is up to the Perl script to parse the string to find the variable names and interpret URL encoding such as %20 for a space.

2.15   Using HTTP POST

By default, form data is URL encoded and sent via an HTTP *GET*. However, this limits the amount of data that can be sent. Most servers limit the length of a URL: a typical limit is 240 characters. Since the limit varies, it can be dangerous to rely on this method unless you know that you are only sending a few, short form fields.

The alternative is to use an HTTP *POST* command. Unlike *GET*, a *POST* can include data as well as a URL. In fact, a *POST* command is very like a server response, it contains an Entity-Header and Entity-Body section. To use *POST*, you must specify the *method="post"* attribute of the *FORM* tag:

```
<FORM action="/cgi-bin/login.pl" method="post">
```

When the data is sent using *POST*, the data is usually sent using in a similar way to *GET*; fields are separated by ampersands and special characters are URL encoded:

```
user=James%20Roberts&pwd=Secret
```

The encoding used to send the field data is specified by the Media Type given in the request header. The typical encoding as shown in the example above is still called URL encoding (even though there is now no URL in the data).

Using *POST* is recommended over GET for sending field data, providing of course that your server supports it, since *POST* does not have the server-dependent length limitation inherent in the *GET* command.

2.16 Submit is a hyperlink

Submitting a form is very similar to clicking on a hyperlink. The browser sends a URL to the server and expects a new page in response — probably HTML, but perhaps something else such as an image. When received, the new page is shown, replacing the previous page (the page which contained the form) in the browser's display.

2.17 Browsing is transaction processing

You can view the interaction between a browser and a Web server as a series of transactions. The interaction can be thought of in the following way.

1. Browser to server: Send me data for this URL.
2. Server sends data.
3. Browser displays data.
(Repeat).

With forms, hyperlinking adds user input data as parameters to the URL.

If you are a programmer or computer scientist, you might like to think of browsing as RPC (Remote Procedure Calls). The URL is the name of the procedure to call. If the URL is sent as a form submit, then the fields are subroutine parameters passed to the remote procedure which is executed on the server.

2.18   Browsing is all pull, no push

HTTP is a client-driven protocol. Everything happens at the initiative of the client (the browser). The server does nothing unless it receives a command from the browser. In Web jargon, HTTP is a pull technology.

Some applications need information from a server when something happens in the world. Perhaps you want to display a big red icon when your inventory falls below a certain level. Or maybe you'd like to see a message when Intel's stock price goes over $100. HTTP and HTML don't support this kind of application, you need a technology which allows the server to send a command, this is called push.

To implement a pull technology you will have to establish an IP connection to the server which is separate from the HTTP session. This might be done by using a Java applet, for example.

2.19   Web servers are stateless

The HTTP protocol is designed to allow each client request to be made in a different TCP/IP session. The session may be kept alive to improve speed, but that's optional. The server is not obligated to store any information about any previous request or previous session with a client, it is up to the browser to remember any needed information.

This design is quite deliberate, it makes it much easier for a Web server to handle large numbers of different clients and for large

Web sites to distribute requests among different physical servers without these servers having to share information. In Web jargon, this is expressed by saying that the Web server is stateless, the term state here refers to information about the session specifically, information that needs to know something about the history of the session.

2.20 Presentation layer and three-tier architecture

Web applications cleanly separate responsibilities between the client browser and the Web server. The browser is responsible for displaying information and collecting user input. The Web server is responsible for processing user input, performing transactions requested by the user and reporting the results in a form which can be displayed by the browser.

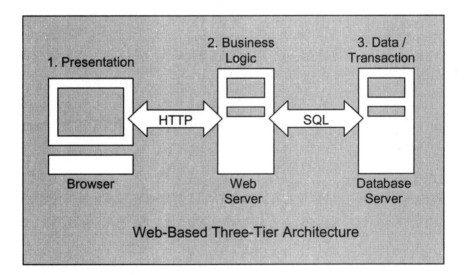

Web-Based Three-Tier Architecture

In this model, the browser is said to be responsible for the presentation layer. This involves a) displaying information in HTML and other formats supported by the browser, and b) collecting user input in the form of clicks on hyperlinks and data entered into forms. In the jargon, displaying HTML etc. is called rendering.

The Web server contains the transaction processing code. If, for example, you want to buy an airline ticket or order a book, the browser will collect the needed parameters (say, by using a form) and send them to the server. The server will run a script which will attempt the transaction. The script is responsible for returning a page which confirms the result of the transaction to the user.

A transaction generally involves executing a related series of queries and updates to a database. The database server is usually separate from the Web server. Scripts will use SQL or some other database access technology to implement the transaction.

Applications built this way are said to have a three-tier architecture. The tiers are:

1. Presentation tier (browser).
2. Business logic tier (Web server).
3. Transaction and data storage tier (SQL server).

3 XML, The Extensible Markup Language

3.1   What is XML?

XML is a framework for creating new languages. An XML
document looks a lot like HTML—not by coincidence, since the
design of XML was heavily influenced by experience with HTML.
It might be more correct to call XML a meta-language rather than
a language because XML allows you to invent your own tags and
attributes and give them any meaning you choose.

As an illustration, we'll invent an XML language which describes
information about people. XML is often used to exchange data
between different software. An older technology to do this is the
comma-separated file. Here is a sample comma-separated file we
will translate into XML:

```
James Roberts,47,(415) 555-1212
Alice Hacker,19,(408) 555-1213
Edwin Eavesdropper,,(212) 555-1214
```

The fields are name, age and phone number. Notice that we don't
know Edwin's age. Here is an XML file containing the same data.

```
<?xml version="1.0"?>
<!-- Information about some people -->
<people>
   <person age="47">
      <name>James Roberts</name>
      <phone>(415) 555-1212</phone>
   </person>
   <person age="19">
      <phone>(415) 555-2212</phone>
      <name>Alice Hacker</name>
   </person>
   <person>
      <name>Edwin Eavesdropper</name>
      <phone>(415) 555-3212</phone>
   </person>
</people>
```

An XML document always starts with a declaration like this:

```
<?xml version="1.0"?>
```

This is a comment:

```
<!-- Information about some people -->
```

The meat of the document is enclosed between a pair of tags
<people>...</people>.

The information relating to a given person is found between
<person> and *</person>*.

We chose to represent the name and phone number fields by
using *<name>* and *<phone>* respectively. Age, on the other hand,
we represented as an attribute of the *<person>* tag. Since we're
inventing our own language here, we're free to use an attribute or
a tag as we please. The main difference is that only one attribute
with a given name may appear within a given tag, but multiple
nested tags with the same name are allowed.

In the comma-separated file, we had to provide the fields in a
fixed order. In our XML file, since the tag names identify the
fields, we're free to put the tags within *<person>...</person>* in any
order we wish. We can also omit tags and/or attributes: notice we
didn't bother to put an *age* attribute for Edwin since we don't
know how old he is.

XML documents are built from the same elements as HTML: tags,
attributes, text and comments.

3.2   XML tags

A tag is written between angle brackets <...>. There are three
kinds of tags: opening tags, closing tags and empty tags.
Technically, tags are called elements, but we believe that the word
"tag" is more easily understood and less likely to be confused with
something else, so that is the word we use in this book.

3.3   Opening tags

An opening tag has the following syntax:

```
<tagname attributes_opt>
```

(The symbol "$_{opt}$" attached to the end of an element means that this element is optional).

For example:

```
<person>
```

Unlike HTML, tag names are case-sensitive: *<Person>* and *<person>* have different names.

An opening tag may include attributes (to be described in more detail shortly), for example:

```
<person age="19">
```

Unlike HTML, it is always illegal to have an opening tag without a matching closing tag.

3.4   Closing tags

A closing tag has the following syntax:

```
</tagname>
```

For example:

```
</person>
```

Closing tags may not contain attributes.

3.5   Empty tags

An empty tag has the following syntax:

```
<tagname attributes_opt/>
```

For example:

```
<phone/>
```

This is exactly equivalent to writing an opening tag immediately followed by a closing tag:

```
<phone></phone>
```

The application processing the XML will treat these in the same way — XML parsers should not allow the application to figure out which of these two forms was used.

An empty tag may contain attributes between the tag name and the slash character, for example if we don't know the person's phone number or name we might write:

```
<person age="29"/>
```

3.6   XML attributes

As with HTML, attributes can appear, following the tag name, in opening tags or in empty tags but not in closing tags.

An attribute is written as *name="value"*.

The value must be enclosed in single quotes '...' or double quotes "...". It is never legal to omit the quotes, unlike HTML.

```
<-- Often accepted in HTML, always illegal XML -->
<FONT size=2>

<-- Correct HTML, strictly enforced in XML -->
<FONT size="2">
```

If double quotes are used, you can include single quotes in the attribute value, and vice versa.

The characters "<" and "&" are always reserved inside attribute values. To represent the single character "&" you can use "&", to represent "<" you can use "<". You can use "'" for a single quote (also called an apostrophe) and """ for a double quote.

3.7 Tag and attribute names

Names must start with a letter, underscore or colon ":", and may not contain whitespace characters (blank, tab or newline). Most punctuation characters are allowed, colon (":") is reserved for namespace features.

User-defined names may not start with the three characters "xml", using upper-, lower- or mixed-case, such names are reserved for the language itself.

XML names are always case sensitive, unlike HTML.

3.8 Comments

An XML comment is like an HTML comment: it starts with the four characters "<!--" and ends with "-->". It is illegal to have two consecutive hyphens within the comment, but otherwise any characters are allowed. Here is a comment:

```
<!-- Here is a legal XML comment. You can
  mention tag names like <person> and use
  special characters including <, >, ", ', &,
  newline and tab - as long as you don't have
  two consecutive hyphens. -->
```

Here is an illegal comment:

```
<!-- WARNING, this is ILLEGAL -- it has two
  consecutive hyphens -->
```

3.9 Parsed text

Most text which is not part of a tag is called parsed text. It is scanned by an XML parser looking for tags, comments and other language elements.

By analogy with HTML, this is considered text which is being "marked up" by the tags.

For example, in the following *James Roberts* is parsed text:

```
<name>James Roberts<name/>
```

Only two characters are reserved in parsed text: the left angle bracket "<" and the ampersand "&". Parsed text can include these characters by using the special sequences "<" for "<" and "&" for "&", which are also valid inside attribute values. These work in the same way in HTML.

Alternatively, you can use a CDATA section. Text inside a CDATA section is not parsed, except to search for the special sequence of characters which is used to end the section. CDATA sections begin with the string <![CDATA[and end with]]>. For example:

```
<text><![CDATA[Text with "<" and "&" but parser
   is not confused]]></text>
```

CDATA sections can contain any characters in any order, except for the "]]>" sequence which ends the section.

CDATA sections may sound like an easy solution, but they must be used used with care. If you have apparently random data (e.g., an encrypted file) embedded in your XML document, or descriptive text which may contain special characters, then it is tempting but risky to put it in a CDATA section without checking for "]]>". What are the odds of this sequence turning up? If it's descriptive text, the odds are increasing every day (because of prose like this which talks about CDATA sections!). If the data is a truly random string of bytes, then the odds are 1 in 256^3 for a random sequence of three bytes, which is 1 in 16.8 million. That may not sound too bad, but that's just three bytes — once you are up to 10 megabytes, the odds are roughly even. Those odds are the same whether you have one large CDATA section or many small sections. How long will it be before your Web site processes 10 megabytes of CDATA sections...?

So let's suppose you do the right thing and scan your data for "]]>". What do you do if you find it? The only alternative is to drop CDATA and use parsed text, which requires you to escape ">" and "&" characters by using ">" and "&". But if you must incorporate that alternative into your application anyway, why not simplify and always use parsed text with escapes?

3.10 Tag values

The value of a tag is what appears between the start tag and the end tag. An empty tag has no value. The value may be a combination of parsed text and/or nested tags. The tag value is often referred to as the content of the tag.

Here, the value or content of the tag ABC is the parsed text *alphabet*:

```
<ABC>alphabet</ABC>
```

In this next example, the value of the tag named *ABC* is a combination of parsed text and two sub-tags (an empty tag named *DEF* and a tag named *GHI*, which itself contains a parsed text value):

```
<ABC>one<DEF/>two<GHI>value of GHI</GHI></ABC>
```

3.11 Overlapping and nesting tags

In XML, as in HTML, tags may be nested. However, unlike HTML, tag values may not overlap. For example, the following is legal in HTML:

```
<!-- Legal HTML but ILLEGAL XML -->
<b>Bold <u>Bold Und </b>Und </u>Normal
```

It would be displayed as:

Bold <u>Bold Und</u> <u>Und</u> Normal

This is not allowed in XML because the tags overlap. To make this legal XML, you could re-write it as:

```
<!-- Legal HTML and XML -->
<b>Bold <u>Bold Und </u></b><u>Und </u>Normal
```

This would display in exactly the same way.

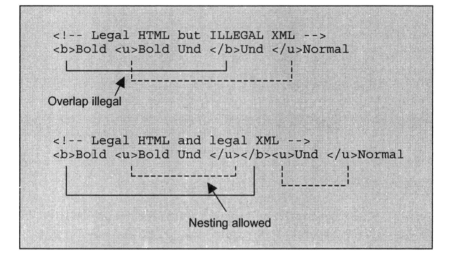

3.12 What do tags mean?

In HTML, there is a pre-defined list of tags and attributes which have fixed interpretations: they describe how to display text in a graphical browser. XML is different, you are free to invent tag and attribute names and meanings as you like.

There is no way in XML to specify the meaning of a tag (except perhaps by using a comment). An XML document is interpreted by an application program, that's what gives it meaning.

3.13 Well formed and valid documents

There are two kinds of XML documents: well formed and valid.

Everything we've presented so far applies to well-formed documents. If you are writing a well-formed document, then you are free to invent tags and attributes as you go along, providing that you obey the syntax rules we've described (e.g., no attributes in a closing tag, no overlapping tag values, and so on).

XML also allows you to specify a fixed language like HTML. You can specify a pre-defined list of tag and attribute names and give precise rules for which attributes can appear in which tags, which tags may be nested in other tags, and so on. The complete set of rules for a fixed language is called a schema, and is specified using a Document Type Definition (DTD). An XML document which is well formed and also complies with a DTD is called valid.

3.14 DTD, the Document Type Definition

Let's return to our original example, an XML document containing information about people. Now we want to specify a schema, a set of rules which govern what should be considered a valid document of this type.

First we'll write down the rules informally in English, then we'll translate to DTD syntax.

> There is a tag named *people*.
> The value of *people* is zero or more *person* tags.

> There is a tag named *person*.
> Person has one optional attribute named *age*.
> The value of *age* is any text.
> The value of *person* is zero or more *name* and *phone* tags in any order.

> There is a tag named *name*.
> The value of *name* is text.

> There is a tag named *phone*.
> The value of *phone* is text.

In DTD terminology, tags are called elements. A tag is defined using the following syntax:

```
<!ELEMENT tagname valuesyntax >
```

The value syntax is specified using a kind of regular expression. Parentheses " (" and ")" are used for grouping. A vertical bar " | " means "or". A star "*" means "zero or more". A plus sign "+" means "one or more". To specify parsed text (known as parsed character data in DTD terminology), use *#PCDATA*. A comma-separated list means that the items must appear in exactly the order written. An empty tag is specified by *EMPTY*.

Following are some examples showing how the value syntax specification works.

Value is parsed text:

```
( #PCDATA )
```

Value is either tag *name* or tag *phone*:

```
( name | phone )
```

Value is either tag *name* then tag *phone*, or tag *name* then tag *address*:

```
( ( name , phone ) | ( name , address ) )
```

Another way to express the same rule is to say *name* then *address* or *phone*:

```
( name , ( address | phone ) )
```

Value is one or more *person* tags:

```
( person )+
```

Value is zero or more *person* tags:

```
( person )*
```

Following are the four tags from our example.

There is a tag named *people*. The value of *people* is zero or more *person* tags:

```
<!ELEMENT people ( person )* >
```

There is a tag named *person*. The value of *person* is zero or more *name* and *phone* tags in any order:

```
<!ELEMENT person ( name | phone )* >
```

There is a tag named *name*. The value of *name* is text:

```
<!ELEMENT name ( #PCDATA ) >
```

There is a tag named *phone*. The value of *phone* is text:

```
<!ELEMENT phone ( #PCDATA ) >
```

Attributes are specified using the following syntax.

```
<!ATTLIST tagname attspec attspec ... >
```

An attribute specification (*attspec*) has the following syntax:

$$attrname\ legalvalues\ defaultvalue_{opt}\ required_{opt}$$

(The symbol "$_{opt}$" attached to the end of an element means that this element is optional).

For example,

```
<!ATTLIST person age CDATA #IMPLIED >
```

For ease of reading, each attribute is often written on a line by itself. Let's add a *weight* attribute, now we might write the definition as:

```
<!ATTLIST person
   age CDATA #IMPLIED
   weight CDATA #IMPLIED >
```

The valid values for an attribute can be given either as a " | "-separated list in parentheses or as *CDATA*, which means text (character data), or as a few other special symbols, to be described shortly.

You might be wondering why we need both *CDATA* and *#PCDATA*, aren't they the same thing? Actually, no; text for an attribute value (*CDATA*) is always found in quotes and follows different parsing rules from a tag value (*#PCDATA*).

For example, the following says that the tag named *play* has one attribute named *mode* which has two possible values, "*tts*" and "*recorded*", and that "*tts*" is the default if not specified:

```
<!ATTLIST play
   mode ( tts | recorded ) tts #IMPLIED >
```

The default is *#IMPLIED*, so this may be omitted. (Strictly speaking, an application could define a difference between *#IMPLIED* and specifying nothing since the parser is required to report which one was used. This is probably not a good idea in most cases since it would be confusing. See the XML specification for details. The VoiceXML DTD always specifies *#IMPLIED* for optional attributes, so this detail does not concern us here.)

You can also omit to specify a default value, so this would also be legal:

```
<!ATTLIST play
   mode ( tts | recorded ) >
```

The application may still define a default, but the DTD should make this explicit if possible.

Note that quotes are optional in the DTD to specify valid values or the default value, but must appear in the document, for example:

```
<play mode="tts">3.14159</value>
```

The following alternatives can be given for the *required* field.

No *required* field given.
(I know this sounds like a contradiction, but the *required* field itself
is optional!). This has essentially the same effect as specifying
#IMPLIED, so by default an attribute is optional. (The application
can distinguish between nothing and *#IMPLIED*, so in theory an
application might react differently).

#IMPLIED
The attribute is optional.

#REQUIRED:
The attribute is required.

#FIXED
The attribute is required and must always be given the (one) value
specified as the default.

Special symbols which may be used to specify a valid value for
the attribute are as follows.

CDATA
Text in single or double quotes.

NMTOKEN
A text string which is a valid XML name. (The name does not
have to appear in the DTD, this means any name which would be
valid for a tag or attribute).

NMTOKENS
A text string which is a whitespace-separated list of one or more
valid XML names.

ID

The attribute is a text string which may be referred to by an *IDREF* attribute elsewhere in the same document. This is similar to the anchor concept in HTML (a URL named something like "#*Top*").

IDREF
A text string which must match an *ID* attribute elsewhere in the same document. Example: If an attribute *goto* is defined as having an *IDREF* value and is specified as *goto*="*Top*", then somewhere else there must be an attribute defined as having an *ID* value which has been assigned the value "*Top*", for example *label*="*Top*".

ENTITY
A tag name from this DTD.

ENTITIES
A whitespace-separated list of tag names from this DTD.

NOTATION
This refers to a <*!NOTATION* ... > definition in the same DTD. This is not used by VoiceXML, see the XML specification for more information.

Now we are ready to show the complete DTD for our language.

```
<!-- Complete DTD for people -->
<!-- Tags: -->
<!ELEMENT people ( person )* >
<!ELEMENT person ( name | phone )* >
<!ELEMENT name ( #PCDATA ) >
<!ELEMENT phone ( #PCDATA ) >

<!-- Attribute: -->
<!ATTLIST person
   age CDATA #IMPLIED >
```

There are two ways to associate a DTD with an XML document: by including the DTD in the document itself, or providing a URL where the DTD can be found.

To provide the DTD in the document you begin with:

```
<!DOCTYPE roottagname [
```

The end of the DTD is marked by:

```
] >
```

The following shows the DTD added to our people sample.

```
<?xml version="1.0"?>

<!DOCTYPE people [
<!ELEMENT people ( person )* >
<!ELEMENT person ( name | phone )* >
<!ELEMENT name ( #PCDATA ) >
<!ELEMENT phone ( #PCDATA ) >
<!ATTLIST person
    age CDATA #IMPLIED >
] >

<people>
   <person age="47">
      <name>James Roberts</name>
      <phone>(415) 555-1212</phone>
   </person>
   <person age="19">
      <phone>(415) 555-2212</phone>
      <name>Alice Hacker</name>
   </person>
   <person>
      <name>Edwin Eavesdropper</name>
      <phone>(415) 555-3212</phone>
   </person>
</people>
```

The other method is to save the DTD in a separate file. You refer
to it using the following syntax:

```
<!DOCTYPE roottagname SYSTEM URL >
```

For example,

```
<!DOCTYPE people SYSTEM "people.dtd" >
```

Whether you include an in-line DTD or refer to a separate file, the *DOCTYPE* declaration should be placed before the XML data which it describes.

DTDs often have repetitive elements. To make DTDs easier to read and write, you may define "entities". An entity is like a *#define* in the C programming language: it allows you to give a string of text a name. When that name is encountered elsewhere, the string of text is substituted in its place.

An entity is defined using the following syntax:

```
<!ENTITY % entityname "text" >
```

The double quotes are removed before the text is substituted. When a percent sign "%" is encountered followed by the entity name, the text in substituted.

Consider the following example from the VoiceXML DTD:

```
<!ENTITY % boolean   "( true | false )" >
```

The entity *%boolean* is used in attribute definitions such as:

```
<!ATTLIST menu
    id      ID      #IMPLIED
    scope     %scope;      'dialog'
    %accept.attrs;
    dtmf      %boolean;      'false' >
```

The last line will be expanded to:

```
    dtmf   ( true | false )   'false'
```

3.15 White space

XML parsers preserve white space (blanks, tabs and newlines) in tag values and inside attribute values. White space anywhere else is discarded by the parser, so it may be used to make the code easier to read. It is up to the application how to process white space in tag values; just because the parser passes it through to the application doesn't mean that the application can't discard it.

The following example of a complete document: illustrates the parsing rules, a white space character is represented by "■" if it is preserved or "□" if it is discarded:

```
<?xml version="1.0"?>
<title□font="Helvetica■Bold">Hello, XML■
Here■we■have■several■
lines</title>□
□□<text>Some■<b>text,</b>■
■■and■so■on.</text>
```

The newline following *</b>* must be preserved because it is part of the value of *<text>*, however the newline following *</title>* will be discarded because it is not part of a tag value.

This document is not typical. Most XML documents have a single tag (called the root tag) which encloses all other tags and values. For example, in VoiceXML the root tag is named *vxml*, so a VoiceXML page is enclosed inside *<vxml> ... </vxml>*. This means that in most XML documents, almost all white space is preserved and passed on to the application. The only white space which is discarded is that found inside a tag but outside an attribute value, as in the following example:

```
<title□font="Helvetica■Bold"□size="10pt">
```

Note that just because white space in a given position is not preserved does not necessarily mean that it can be omitted: you must, for example, have at least one white space character between a tag name and an attribute name.

Note also that even though the parser preserves white space, the application may re-format the text in any way it chooses. A particular XML language like VoiceXML may have additional specifications which require it to treat white space in a certain way.

There is a special (and often misunderstood) attribute named *xml:space* which is reserved by the XML specification to assist with white space processing. If used, it must be assigned either *"default"* or *"preserve"*. If set to *"preserve"*, this signals the application that the author of the document intended white space in the value of that tag to be preserved. If set to *"default"*, this means that the author of the document accepts the default behavior of the application (whatever that might be, the XML specification leaves that up to the application to decide). Note that even if *xml:space* is set to *"default"*, the XML parser will pass all the white space through to the application. Don't believe the common claim that white space will be discarded by the parser, though it may be discarded by a particular application. A child tag will inherit the value from its parent if it is not specified; the parser takes care of this so that the application doesn't have to deal with it. This means that you can specify a value in the root tag and this will set a default for the whole document. If you use *xml:space* and are validating against a DTD, then the DTD must explicitly specify *xml:space*.

VoiceXML does not use *xml:space*; white space is therefore always passed on to the browser.

3.16 HTML and XML compared

For those readers who are familiar with HTML but new to XML, the following table gives a summary of some of the most important similarities and differences between the two languages.

Feature	HTML	XML
Document consists of tags, attributes and text.	Yes.	Yes.
Names of, and interpretation of, tags and attributes.	Pre-defined.	Defined by the user.
Tag and attribute names are case sensitive?	No.	Yes.

Feature	HTML	XML
Open tags allowed without closing tags?	Sometimes.	Never.
Attribute values allowed without quotes?	Sometimes.	Never.
Parser should try to handle documents with incorrect syntax?	Yes.	No.
White space preserved?	Usually not.	Always preserved by parser, up to application to process.
Reserved characters?	Several.	Only "<" and "&", plus either single or double quote inside an attribute value.
Nested tags allowed?	Yes.	Yes.
Overlapping tags allowed?	Sometimes.	Never.

3.17   The XML document tree

An XML document can be viewed as having a tree structure.
Consider this example VoiceXML document.

```
<vxml version="1.0">
   <form>
      <field name="drink">
         <prompt>
            Coffee,tea, or milk?
         </prompt>
         <grammar src="drink.gram"/>
      </field>
      <block>
         <submit next="drink2.asp"/>
      </block>
   </form>
</vxml>
```

A tree representation of this document could look like the following diagram.

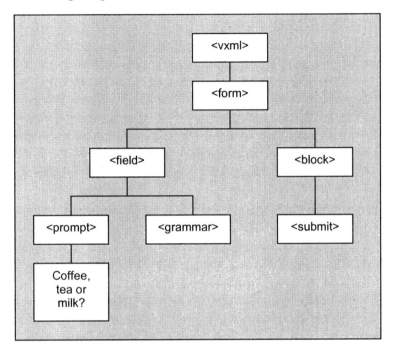

Viewing the document as a tree is sometimes helpful when you want to understand the relationship between tags. We can describe the *<promp>* tag as a child of *<field>*, for example, or say that *<block>* is a parent of *<submit>*. The sub-tree under a given tag is the value (content) of that tag.

4 Telecommunications

4.1   Introduction to telephony

In this chapter, we assume you're familiar with how to use a
telephone, but know little about telecommunications technology.

There are many different types of phone line, but we'll illustrate as
many concepts as possible with Plain Old Telephone Service
(POTS), the kind of service you probably have at home.

A POTS line is a loop of wire which starts at a local telephone
company Central Office (CO) switching station. The CO switch
applies a small amount of power to the line, this is called battery
voltage, or just battery. Your telephone has a small on/off switch
(not to be confused with the CO switching equipment) called a
hook switch. When the telephone is in its resting position, the
hook switch is off and there is no current flowing; the phone is
said to be on the hook, or simply on-hook. When you pick up the
telephone, this closes the hook switch (turns it on), completing the
circuit and causing current to start flowing. Since this current is
flowing along the local loop between your phone and the CO
switch, this current is called loop current.

The local loop is sometimes referred to as the last mile because the
distance is typically within a mile or two. It is the most
troublesome link in the telecommunications network because
even though most central offices are now based on sophisticated
digital equipment, the last mile and the customer's phone are
generally based on this antiquated analog technology.

If you look inside a standard telephone cable, you will see that
there are two wires, this kind of cable is called twisted-pair
(because the conductors are twisted together to reduce
interference effects). The wires meet at each end of the line: the
CO switch and the hook-switch inside your phone, creating a
complete circuit.

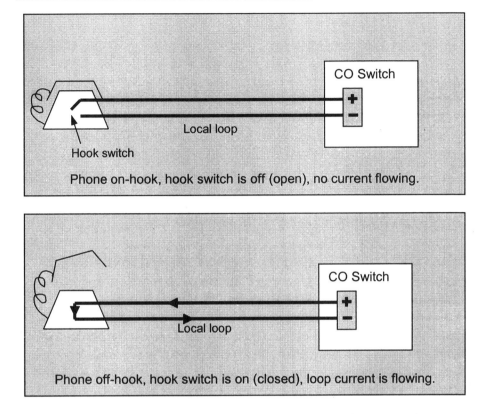

Phone on-hook, hook switch is off (open), no current flowing.

Phone off-hook, hook switch is on (closed), loop current is flowing.

Sound is carried by varying the current flowing in the wire. The microphone in the telephone varies the resistance according to the volume of the source: when the volume is high, the resistance is low and more current flows; when the volume is low, the resistance is high and less current flows. The amount of current flowing therefore indicates the level of the sound. The loudspeaker in the earpiece does the reverse translation: the current flow is converted into movement of a membrane, which causes the air to vibrate, which you hear as sound. There is only one wire which carries both your voice and the voice of the person you are speaking with, so in fact you hear yourself just as much as the other party. To convince yourself that this is true, try blowing into the microphone, you'll hear a gale force wind in the earpiece; this doesn't happen on modern digital phones which plug directly into an ISDN line because digital lines have separate "talk paths" for your voice and the other party's voice.

The current that flows during a normal conversation is DC (direct current). To ring the phone, the CO switch applies a higher-voltage AC (alternating current) voltage to the line (it's enough that you'll feel a distinct tingle if you are holding onto a bare telephone line when the switch tries to ring it). The phone is able to detect the AC because there is a capacitor which makes a complete (high-resistance) circuit even when the hook switch is off. (The resistance is high enough that almost no current flows, but it is possible to detect it with specialized equipment—that's you can eavesdrop on a room using a phone even when it's on-hook). The capacitor responds to the AC, ringing the bell.

To dial a number you must send some digits to the CO switch. There are two dialing systems used by POTS: touch-tone and pulse, also called rotary or (rarely) decadic dialing.

Touch-tones send digits through sound. Pressing a digit button sends a pre-defined sound composed of two pure frequencies. Each digit sends a different pair of frequencies. For this reason, touch-tone digits are often referred to as DTMF, for Dual Tone Multi-Frequency.

Pulse digits are sent by turning the loop current rapidly on and off. The number of pulses (a pulse is one current on-off-on cycle) indicates the digit. Zero is specified by 10 pulses. Pulse dialing is also called rotary because it is the system used by the older rotary dial phones (which was the only kind of phone back in the days when dialing actually involved a dial). Pulse dialing is slower than tone dialing and is very hard for automated equipment to deal with because it's only visible to the first switch to which the phone is connected: all subsequent devices in the call hear only clicks, the interruptions in the loop current are not passed on (for the excellent reason that later steps in the chain don't use analog signaling technology).

4.2 Life cycle of the common phone call

There are two types of phone calls: in-bound calls, in other words calls where your phone rings and you answer, and out-bound calls, which you initiate by dialing a number.

Out-bound calls are established by the following sequence.

1. Seizing the line. This is telecom jargon for picking up the phone in order to request service. The switch signals that it has detected your seize and is ready to receive digits by playing dial tone. Dial tone is another dual tone, a tone comprised of two pure tones (different from the dual tones sent by touch-tone digits).

2. Dialing. You send digits to the switch by sending DTMF tones or pulses.

3. Call progress. The phone network sends you sounds which indicate how your call is progressing. For example, you might here a busy tone which indicates that the line you are calling was already off-hook. You might hear ringing tones which confirm that the switch attached to the called phone is currently ringing the line (ringing tones are produced by the called switch, not by the called phone—there could be no phone at all attached to the line and you would still hear ringing). Or you might hear a message like "that is an invalid number, please hang up and try again", or "this number has been changed, the new number is...". Call progress signaling (these tones and messages) are designed for people, not for computers—it is not at all easy to create automated systems which can dial a call and determine whether the call was connected successfully. Call progress analysis algorithms are rarely 100% accurate, especially when dialing numbers which may use different tones and messages (e.g., dialing locally, long-distance and internationally to different countries).

4. Call completion. If the called party picks up their phone, the call is put through, or "completed" in telecom terminology.

At any time during this sequence, you can of course abort by simply hanging up the phone.

In-bound calls are much simpler. They proceed as follows.

1. Ringing. You hear the phone ring. This line state is called offering, meaning that a call is being offered to you.

2. Call completion. You pick up the phone, completing the call.

While this sounds simple enough, there is a wrinkle here that you may never have noticed: all phone lines which support both in-bound and out-bound calls suffer from a potential problem. If you pick up the phone to make an out-bound call when the line is about to ring with an in-bound call, you will end up answering that call instead of getting dial tone. This phenomenon is called glare.

When the call is completed, a talk-path is established in both directions so that the two parties can hear each other.

Either party can end the call by hanging up the phone (putting the phone back on-hook). In the United States, if you are connected to the public network via an analog line and the other party hangs up, you are notified of this in the following way. The CO switch will turn off battery voltage for one second, causing a one-second drop in loop current. This so-called loop drop disconnect supervision is for the benefit of automated equipment, a person listening to a typical phone has no way to monitor the current. Unfortunately, many company phone systems (PBXs, for Private Branch Exchanges), and public phone lines in many countries, do not provide this notification. In some countries you will hear a special tone, similar to a dial tone or busy tone, to indicate that the other party hung up. Sometimes you will get no notification at all. As with call progress analysis, this can be a challenge for computer equipment (such as a VoiceXML browser) which needs to know when the caller hung up. Fortunately, digital lines are available in most locations and these generally give some form of disconnect notification.

4.3   PSTN, The Public Switched Telephone Network

The global telephone network is known as the PSTN, for Public Switched Telephone Network.

As we have seen, your home phone is connected to a CO switch. The capacity of such a switch is typically something like 10,000 lines. In the US, a local phone number has seven digits. The first three digits are often used to identify which switch is attached to the line, the last four digits identify the line number on that switch. So, if your number is 555-5656, and your neighbor's number is 555-1234, chances are that your lines are attached to the same switch. Each switch has a set of routing tables so that it knows how to proceed with a given call. The role of these tables is similar to those used by an IP network router.

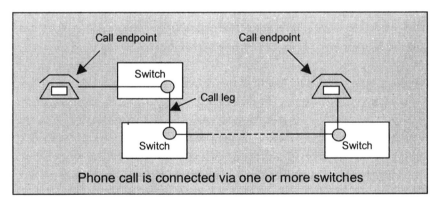

Phone call is connected via one or more switches

If the dialed number belongs to a line on the same switch, it can complete the call by connecting the two lines and no further links are needed. If it is a long-distance or international call, the switch must identify the carrier and forward the call to an outgoing connection which reaches that carrier's closest switch. Analogous to the Internet backbone is a set of high capacity, high-speed switches at the heart of the long-distance and international networks.

The function of a switch is to accept a request for a call and, add a link ("call leg") to the chain of connections for the call, and forward a similar request to the next link in the chain. That next link could be the called party or could be another switch. The process of extending a request for a call from one switch to the next is very similar to the request from the original caller — setting up a call leg is like dialing a number from the originating endpoint of the call. Additional information (ANI, account information for billing) will be passed along together with the number originally dialed. The phone network has its own protocols which take care of storing the routing tables, storing billing information, setting up legs between switches, and so on. One commonly used protocol is called SS7, for Signaling System 7.

4.4   Phone Line Features

In addition to making and accepting calls, there are many other features that a phone line can offer. Some of the more common services are as follows.

4.4.1   Caller ID

You are probably familiar with Caller ID, which is available on many home phone lines. This service provides the phone number and/or name of the caller. It is sent on standard analog lines.

4.4.2   ANI

ANI (Automatic Number Identification) is similar to Caller ID. Many digital lines provide the option of ANI service. Unlike Caller ID, ANI usually provides only the caller's number, not the name. For some applications, the ANI may function as a PIN code and be used to identify a subscriber to your service or an account number. As with a browser cookie, to use ANI in this way you have to assume that this phone is used by only one customer.

4.4.3   DNIS

DNIS (Dialed Number Identification Service) sends the number that the caller dialed. This allows you to have many different phone numbers assigned to the same trunk. The equipment that answers the call can check the DNIS and answer the call appropriately.

DNIS is used by organizations in order to provide direct dial lines through a PBX. The PBX answers a call, looks up the local extension assigned to the DNIS, and rings that line.

It is also used by automated systems to determine which service to provide. For example, a customer care call center serving different companies can answer the call using a computer. The computer (say, a VoiceXML browser) determines which script to run based on the DNIS, for example a call might be answered "Thank you for calling company A, please hold for an agent" or "Welcome to service B, please enter your PIN code".

4.4.4   Three-way calling and call transfer

Three-way calling allows you to add a third party to a call which is already in progress. On a home phone, you do this by issuing a flash-hook, which causes the hook switch to close for a short time (about half a second, not long enough to hang up the call). Your phone probably has a button marked Flash to do this, if not you do it by quickly depressing the hook switch. The CO switch responds by putting the other party on hold and giving you a new dial tone. You dial the new number, complete the call, flash-hook again and now the three of you will be talking (provided the original caller didn't hang up in the mean time!).

Call transfer allows you to make a three-way call and then hang up without disconnecting the other two parties, leaving the two of them in a one-on-one conversation. You may be able to make a blind transfer, also known as an unsupervised transfer, where you hang up without waiting for the third party to answer the call, which leaves the second party listening to the third party's phone ringing (or other call progress information, such as a busy tone or error message).

4.4.5   Toll-free and premium billed services

A toll-free service is billed to the called party instead of the caller. In the US, toll-free services are assigned special area codes, most commonly 800, plus new codes such as 888 and 877 which are now coming into use.

A premium service charges extra for a call. In the US, area code 900 is used for such services. The charges appear on the caller's phone bill, which the phone company collects and pays to the owner of the 900 number (minus phone company fees). After a surge of popularity in the late 1980s and early 1990s, premium services fell into disfavor after widespread abuse and complaints; many phone companies no longer offer this type of service. They are still used for a few applications, notably for paid software support.

4.5   Digital audio

We've seen that analog phone lines carry sound by varying the amount of current flowing. Sound is represented digitally by a sequence of numbers. Each number, called a sample, represents the loudness of the sound at a particular time. A rapid series of samples can give an accurate approximation to a given sound. Digital sound is carried by digital phone lines and stored in sound files, such as Wave or MP3 files, and on music CDs.

Digital audio is characterized by a number of parameters including the sampling rate, bits per sample, number of channels, companding and compression scheme.

4.5.1   Sampling rate

A music CD samples sound at 44,100 samples per second (44.1 kHz). The public telephone network uses 8,000 samples per second (8 kHz). Sound files for use on computer telephony equipment often use an even slower rate, 6 kHz, to reduce the load on the hard drive and operating system and allow larger numbers of lines to be supported on a single computer.

4.5.2   Bits per sample

A music CD uses 16-bit samples. The public telephone network uses 8-bit samples. Computer telephony sound files typically use 4-bit samples (to minimize load) or 8-bit samples (for best quality).

4.5.3   Number of channels

A music CD uses two channels, in other words, stereo. In telephony, one channel is almost always used for obvious reasons.

4.5.4   Companding

Companding is a technique which squeezes better sound quality out of a given number of bits per sample. The human ear is more sensitive to quiet sounds, so if you use a non-linear translation between the numerical value and the amplitude of the sound to give more detail at the quiet end of the spectrum at the expense of less accurate louder sounds, the result may sound better. Most audio data carried by the telephone network is companded. In the North America a scheme called mu-law or µ-law is used, in most other countries a similar but different scheme called A-law is used. Data without companding is referred to as linear.

4.5.5   Compression

Data compression is used to reduce the number of bits per sample. In telephony, the most widely used compression is called ADPCM, for Adaptive Pulse Code Modulation. Uncompressed data is called PCM (Pulse Code Modulated). There are many different types of ADPCM. Telephony expansion cards from Intel's Dialogic division often use a variant called Dialogic/Oki ADPCM, which compresses linear 8-bit samples down to 4 bits. ADPCM is a "lossy" compression scheme, which means that the data may not be faithfully restored by a compression followed by a de-compression. Using ADPCM usually degrades the perceived quality of the sound by a small but noticeable amount.

4.5.6   Sound Files

A sound file, as found for example on a desktop computer, is characterized by the digital audio format used plus the structure of the file itself.

A Wave file for Windows has a so-called RIFF header (standard for multi-media files in Windows) followed by a number of "data chunks", one of which (the Wave chunk) will contain the audio data.

A "raw file", such as the commonly-used Dialogic Vox file format, contains only audio data; there is other no data such as a file header stored in the file. This is simple and fast, but has a major drawback: there is no way to determine automatically the parameters of the audio data (bits per sample, compression etc.). The application must use a format which is fixed in advance or must invent a proprietary scheme to keep track of what is stored in the file (e.g., by using different filename extensions).

Some specialized sound files contain several audio recordings. This is used, for example, to store the large number of short recordings used to synthesize messages which include variable values, such as "You have twenty one new messages", or "Your bank balance on January first was nineteen dollars". One such file type in common use is variously called VAP (Vox Array of Prompts), IPF (Indexed Prompt File), or VBase40 (after a defunct sound file editor of that name).

4.5.7   PSTN Digital Audio

The public telephone network carries audio data using 8 kHz, 8-bit, companded PCM. In North America, the companding scheme is mu-law, in most other countries it is A-law. A single channel therefore carries 8 kilobytes per second, or 64 kilobits per second (kbps).

4.6   Phone Line Types

Following are some of the more common types of phone line. A phone line is sometimes called a trunk, especially when it carries a multi-channel digital service.

4.6.1 Loop Start

Another name for the vanilla line type you probably have at home is loop start or simply analog.

4.6.2 T1

Also called DS1 (Digital Service level 1), T1 is a digital trunk which carries 24 channels. It uses two twisted-pair cables exactly like analog lines. Each pair carries 24 channels, one pair carries audio from the customer to the phone company, the other pair carries audio from the phone company to the customer. A single channel is called a DS0 or a time-slot, one channel carries 64 kbps. There are two common variants of T1: robbed-bit and PRI ISDN.

Robbed-bit T1 uses all 24 channels for telephone calls. Signaling information is carried "in-band", i.e. using the same channel that carries audio. Dialing, ANI and DNIS use DTMF digits (or in some cases another kind of dual tone digit signaling called MF or R1). Seize, ring and hang-up are signaled using "robbed bits", where the least significant bit of every 16th sample of the audio channel is used to indicate that the connection is enabled (bit is 1) or disabled (bit is 0). Stealing this bit causes a tiny degradation of the audio, but this is imperceptible.

PRI ISDN (Primary Rate Interface to the Integrated Services Digital Network) uses 23 channels for calls and one channel for signaling. This is called "out-of-band" signaling and "clear-channel audio" since the audio channels are used only for carrying sound. The audio channels are called B (bearer) channels, the data channel is called the D channel. PRI ISDN is sometimes referred to as 23B+D. The D channel carries digital packets where bits represent data directly. This is much faster and more reliable than using audio tones, at the expense of losing one channel per trunk. Using a service called NFAS, multiple T1s can share a single D channel, so it is possible to achieve almost the same number of channels as robbed-bit.

Robbed-bit is fairly well standardized, however PRI ISDN has a number of different flavors. Either way, hooking up equipment to

T1 may involve setting a number of different parameters to match the trunk protocol.

4.6.3   E1

Outside North America, this is the most common digital trunk. It carries 32 channels. Two of these channels are always devoted to signaling, the remaining 30 carry clear-channel audio. E1 can therefore be called 30B+2D. There are two common variants of E1: CAS and PRI ISDN.

E1 CAS (Channel Associated Signaling) is a strange hybrid: the D channels carry signaling bits which take care of seize, ring and disconnect, the B channels carry in-band tones for dialing, ANI and DNIS. How this works in detail unfortunately varies a great deal from country to country or even within a given country. Several tone signaling protocols are used: DTMF and R1 (a.k.a. MF) is sometimes found as for T1, but these are rarely used; alternative but more complex dual-tone schemes called R2 and Socotel are in wide use. If you are hooking equipment up to an E1 trunk, tuning the protocol to match the service on the trunk can be a challenge.

PRI ISDN (and variants called things like Euro-ISDN) is also found on E1. This works in a similar way to PRI ISDN on T1, though again the details vary.

4.6.4   T3

The next step above T1 or E1 is usually to go to T3, which carries 576 channels (equivalent to 24 T1s or 18 E1s) in a single digital trunk. Also called DS3.

4.6.5   BRI ISDN

BRI, or Basic Rate Interface, is a 2B+D flavor of ISDN. In other words, it carries three channels, two of which are used for conversations, one for signaling. Since BRI can carry no more than two calls, it is mainly targeted at home users.

4.7   IP telephony

So far we've discussed traditional types of telephone lines. IP networks, such as the Internet, can also carry voice conversations. This technology is referred to as IP Telephony, Voice over IP (VoIP) or Internet Telephony.

There are two significant protocols which allow two computers to establish a telephone conversation over an IP network: H.323 and SIP (Session Initiation Protocol). SIP is newer and simpler, and appears to be gaining in popularity; H.323 is more widely adopted at the time of writing (January 2001). The basic call control signaling in H.323 (seize, dial, complete, disconnect) is based on Q.931, which is the International Telecommunications Union (ITU) standard for digital signaling using an ISDN D channel.

To save bandwidth, H.323 offers the G.723.1 compression scheme which can transmit with acceptable (though not great) voice quality at 6.4 kbps or 5.3 kbps (that's less than 10% the data rate of a DS0 channel as used on T1 or E1). A major drawback of G.723.1 compression for automated equipment such as a VoiceXML browser is that DTMF tones are distorted so much that they are not recognizable.

Alternatively, H.323 allows G.711 coding, which is exactly the 8-bit, 8 kHz mu-law or A-law PCM encoding used by the telephone network (64 kbps per channel) and will (assuming no packet loss or delays due to network latency) achieve the same voice quality as the PSTN.

If two computers on the same network are calling each other, H.323 or SIP may be all you need. If you want to complete a call involving one or more parties on a traditional phone network, you need a device to convert between IP packets and old-style trunks. Such a device is called a Media Gateway, or just a Gateway.

A Gateway can work around the DTMF transmission problem. Remember that a caller's touch-tone digits, which provide user input for menus, entering credit card numbers and so on, cannot be transmitted in-band when using highly compressed audio such as G.732.1. The Gateway can do DTMF detection and transmit the digits out-of-band. Extensions to H.323 have been proposed to standardize how this is done.

5 Computer Telephony

5.1   Computers and the telephone

The generic term Computer Telephony (CT) is used to refer to any situation where a computer is involved in handling a telephone call. Today, that means almost every call you make: most switches on the public telephone network are controlled by computers, often running the Unix operating system.

Usually, "Computer Telephony" is reserved for configurations where a general-purpose computer is used. Many enterprise telephone systems (PBXs) are controlled by microprocessors but, unless they are based on a general-purpose architecture such as a PC, these are generally considered embedded systems rather than CT systems.

5.2   Computer Telephony Integration

While Computer Telephony Integration (CTI) may sound just as generic as "Computer Telephony", it is usually used to refer to a more specific situation where a switch is controlled by a physically separate computer. The computer and switch are connected via a CTI link, which could be an RS-232 serial line, Ethernet or other type of network connection.

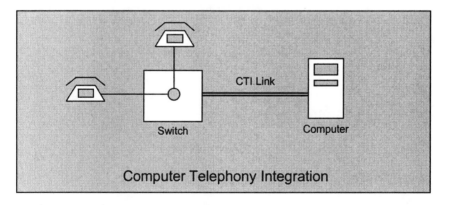

Computer Telephony Integration

The CTI link carries a) messages from the switch to the computer, and b) commands from the computer to the switch.

Typical messages notify the computer of incoming calls and of caller hang-ups.

Typical commands instruct the switch to answer a call, transfer a call to an outside line or a local extension and to set up a conference with two or more parties.

Switches which support CTI are sometimes called smart switches or programmable switches.

5.3   Voice modem

A voice modem is a single-line PC data modem which has been enhanced to support playing and recording messages and detection of touch-tones. Voice modems hook directly to an analog phone line, typically a PSTN line or PBX extension.

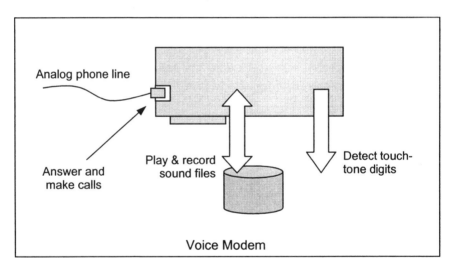

The most common programming interface for voice modems under Windows is TAPI (Telephony Applications Programming Interface), with Wave or DirectX used to play and record.

Voice modems are often inexpensive, but they can have a number of disadvantages. Sound quality can be poor and only a limited range of sound file types may be supported. Hang-up detection can be unreliable or non-existent, especially when the modem is hooked up to a typical PBX which uses tones rather than loop current supervision for hang-up. Since voice modems usually support only one line per expansion card, configurations are limited to at most a few lines per PC, depending on the available slots and the capabilities of the drivers. Call progress detection is usually not supported, in which case the application will be unable to determine if an out-bound call was answered.

Adequate for simple applications such as home voice mail, voice modems are generally unsuitable for professional applications.

5.4   High-end voice cards

Professional PC expansion cards are dedicated to voice processing, they rarely support data modem features. (Hence the name "voice card" rather than "voice modem").

They generally feature one or more high-power DSPs (Digital Signal Processors), specialized microprocessors which are designed to process digitized signals such as audio and video efficiently. Professional voice cards offer better audio quality, large numbers of phone lines and high-quality algorithms for call progress analysis, audio compression / decompression, echo cancellation and other features.

Voice cards usually also support one or more types of voice bus, a high-speed, real-time bus which can be used to connect two or more voice cards together. With a voice bus, a PC gains the ability to connect two or more calls and can therefore function as a switch.

Voice card features can be divided into call control, media processing and enhanced speech technologies. There is some overlap in these categories, as we shall see.

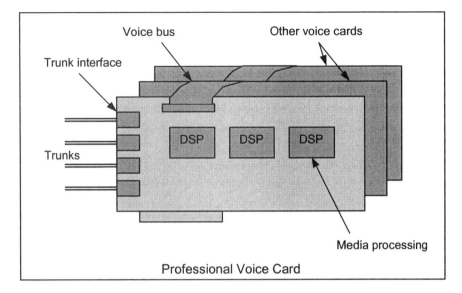

Professional Voice Card

The major features offered by voice cards are as follows.

Category	Features
Call Control	Trunk interface Call bridging
Media Processing	Tone detection Tone generation Media player Media recorder
Enhanced Speech Technologies	Voice recognition Speech synthesis

A leading manufacturer of professional voice cards is Dialogic Corp., an Intel company (the author's employer).

5.5   Voice Bus

A voice bus carries real-time data between components on a single voice card and/or between different voice cards. This is done using Time Division Multiplexing (TDM).

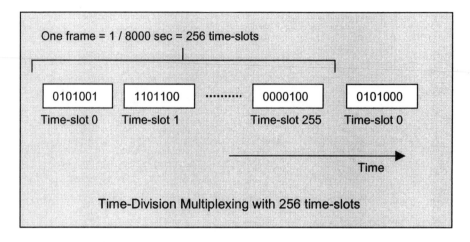

One frame = 1 / 8000 sec = 256 time-slots

| 0101001 | 1101100 | ·········· | 0000100 | 0101000 |

Time-slot 0   Time-slot 1   Time-slot 255   Time-slot 0

Time

Time-Division Multiplexing with 256 time-slots

TDM combines *n* bit-streams into one stream by using a data rate *n* times as great, or by carrying samples *n* times as big at the same data rate, which ever way you prefer to think about it. For example, consider a TDM bus carrying 256 channels where each channel carries 8,000 8-bit samples per second. These channels can be combined to make one channel with a sample size of 256 x 8 = 2,048 bits, at the same rate. In other words, 256 channels at 64 kbps are combined into one channel at 256 x 64 kbps = 16.384 Mbps. A combined sample, containing one 8-bit sample from each channel, is called a frame. A single channel carried by TDM is called a time-slot. The time-slot number (0 to 255 in this example) identifies the position of the channel within the frame.

Digital trunks, such as T1, E1 and SS7, use Time Division Multiplexing technology which is very similar to that used by a voice bus.

The process of transmitting audio data from one device to another over a TDM bus is purely electronic and therefore almost instantaneous — there is no software processing such as routing table lookup, packetization / de-packetization, encryption or decryption, checksum calculation, stripping of packet headers, acknowledgement and so on. Alternative architectures use IP networking instead of TDM, these may suffer from higher latency and other issues (packet loss, jitter and so on).

What exactly is a "device" on a voice bus? By definition, it is a component that produces or consumes a single 64 kbps bit stream. It depends on the design of the card exactly what features are consolidated into different kinds of device. In the case of current Dialogic boards, devices mostly fall into the following three types.

1. Trunk interface.
2. Media processing (player / recorder and tone detector / generator).
3. Conferencing.

Bus devices are sometimes called resources.

A TDM bus functions as a switch, in other words as a means to transmit data between any two devices connected to the bus. When a connection is made via a voice bus, one device (the receiver or listener) is said to be listening to the other (the transmitter). This is called a half-duplex connection. If the two devices are listening to each other, this is called a full-duplex connection.

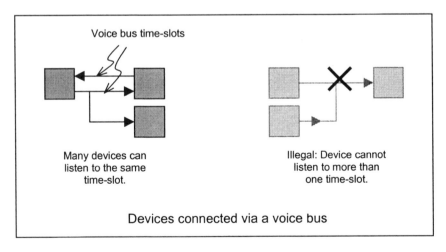

Voice bus time-slots

Many devices can listen to the same time-slot.

Illegal: Device cannot listen to more than one time-slot.

Devices connected via a voice bus

A given time-slot is set by one device, or is all-zeros because there is no device transmitting to it. If two devices attempt to transmit to a TDM bus at exactly the same time (i.e., to the same time-slot), then at best the bit values will be ambiguous, or at worst electrical damage to the bus hardware will occur. Each device (trunk interface, media player etc.) is therefore assigned a time-slot number which is unique among all devices which are attached to the bus. However, any number of devices can receive data from a time-slot since listening is a purely passive process.

5.6   Trunk interface

Current voice cards (January 2001) typically offer up to 16 analog trunks or four digital trunks (T1 or E1) per card. With four E1s, that makes 4 x 30 = 120 lines on a single card.

For Voice-over-IP (VoIP) applications, specialized IP network cards (which connect to Ethernet or other high-speed IP LANs) are available which support IP-based protocols such as H.323.

The primary job of a trunk interface is basic call control: detecting and accepting in-bound calls, making out-bound calls and detecting caller hang-up.

Some types of trunk, including analog, robbed-bit T1 and E1 with R1 or R2 signaling, use tones to dial out-bound calls and to transmit ANI and DNIS digits for in-bound calls. Analog trunks may additionally need specialized tone detection in order to detect hang-up tones which are generated by some PBXs in place of loop drop supervision. Interfaces for these trunk types may have built-in tone detectors and generators, or may need to "borrow" media processing resources from another module on the same card or from a different card. All-digital trunks, such as ISDN and SS7, do not need tone detection or generation for basic call control.

For out-bound dialing, basic call control may not be enough: applications will usually require call progress analysis, which attempts to determine whether the call was answered. In most cases, even all-digital trunks such as ISDN do not provide call progress supervision through the out-of-band signaling, the voice card must attempt to analyze call progress tones. Call progress analysis may be considered as a media processing feature, and may be supported by a media module rather than a trunk interface module. As with tone detection and generation, performing call progress may therefore require reserving a media module and providing it with audio received from the trunk.

Once a call is completed, the trunk interface manages two audio streams. One audio stream is transmitted to the caller, which will either be silence or will be generated by one of the other modules such as a media player. The other audio stream is the caller's voice, which is forwarded by the trunk to other modules which may request to monitor and analyze it. On digital trunks (such as T1, E1 and SS7), these audio streams will simply be copied to and from the trunk as-is, or may be translated to a different format used by other components. For example, a media processing card may expect the audio to use mu-law companding, which is standard in North America. An E1 trunk interface module might convert A-law on the E1 to and from mu-law companding used in the rest of the system. On analog trunks, the trunk interface will be responsible for the analog to digital and digital to analog conversions.

5.7 Call bridging

If the card has a voice bus, then it should be able to support call bridging by using the bus to connect the audio between the two calls, as shown in the following diagram.

Note that because there can be only one source of audio for a given time-slot, the trunk interface can either transmit sound from the other trunk, or sound from a media resource such as a player, but not both. This means that if you want to play a message to a caller in a bridged call, you must first disconnect audio from the other caller (unless you use conferencing to create a three-way call between the two callers and the player).

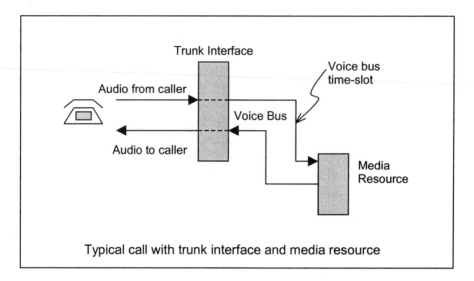

Typical call with trunk interface and media resource

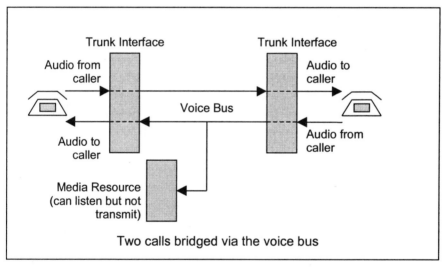

Two calls bridged via the voice bus

5.8 Tone detection and generation

Some of the tones more commonly used by telephony systems are listed in the following table.

Tone Type	Description
DTMF (touch-tones)	Used for dialing numbers on analog and robbed-bit T1 trunks. User input for "voice menu" selections, entering credit card numbers, zip codes etc. Also used to transmit ANI and DNIS on some robbed-bit T1 trunks.
Call progress tones	These include number busy, network busy (also called fast busy), ringing tones, error tones (which come before spoken messages such as "number unobtainable").
Dial tone	Indicates that the switch is ready for dialing.
R2 and Socotel	Used for dialing numbers, also for ANI and DNIS, on E1 CAS trunks.
Hang-up tones	Used by some PBXs and switches to signal caller hang-up.

Most of these tones are dual tones (meaning they are composed of two pure sine waves at different frequencies), a few are single tones (one frequency). They may have a cadence, meaning that there are regular cycles of sound separated by silence (US busy tone is an example), or continuous (such as US dial tone).

Generating tones is straightforward; DSP code can calculate the numerical values for the sine waves, or use a table of pre-stored tables containing the sample values for one complete cycle of the tone.

The detection of tones requires analyzing the received audio and breaking it down into frequency components. This is done by using algorithms called Fast Fourier transforms.

Detecting touch-tones suffers from two potential types of false-positive detection called talk-off and play-off.

Talk-off happens when the caller is speaking and for a short moment the sound of his or her voice matches the frequencies of a

DTMF tone. This happens most often with women whose voices tend to match tone-tone frequencies more often than men. A typical symptom is that the caller is leaving a voice mail message but is cut off short because the system incorrectly detects a tone, terminates the recording and moves on to the next menu. Talk-off can be reduced by tuning parameters for DTMF detection: to require a longer time for the frequencies to be present, setting a lower minimum for other frequencies present (background noise threshold), and so on. However, this may tend to cause more false-negative errors, in other words, failure to detect when the user did in fact dial a tone.

Play-off is a similar phenomenon that happens when a prompt being played by the computer contains a brief period matching a touch-tone. (Remember, when the caller has an analog phone, the audio received from the caller also includes the audio transmitted by the computer which has "looped around" the caller's analog line; the computer is therefore listening to itself as well as the caller). Play-off can be corrected by echo cancellation, where incoming audio is analyzed to subtract the prompt being played (not so easy, since the prompt will be slightly delayed and distorted), or by pre-filtering the pre-recorded prompts used by your application to make sure that they do not contain DTMF frequencies.

5.9   Media players and recorders

A media player copies digital audio to the voice bus (or directly to a trunk interface or other component). The audio may originate in a sound file, a speech synthesizer or other source.

A media recorder copies digital audio from the voice bus (or directly from a trunk interface or other component) to a destination. That destination might be a sound file or a software resource such as a speech recognizer.

Media players and recorders provide buffering services so that the source or destination does not need to be exactly synchronized with the bus. They may also provide format conversion, for example to support sound files which use a different encoding than the voice bus.

5.10   Enhanced media processing

The term enhanced media processing refers to voice recognition and speech synthesis (text-to-speech). These are sophisticated technologies which receive extensive support in VoiceXML. They are covered in detail in another chapter.

In older systems, enhanced media processing was typically done using special-purpose voice cards such as Dialogic's Antares, which provided high-power DSPs for the CPU-intensive algorithms which are needed for these algorithms. In modern systems, the host PC's CPU is often used — the processing power of CPUs continues to increase according to Moore's Law, and it is now not unusual to find 48 or 96-channel enhanced media processing systems based on a single PC chassis.

In fact, all the algorithms we have been discussing — tone detection and generation, analyzing call control signals, playing, recording and so on — are candidates for moving from specialized, DSP-based cards to the host CPU. As the power of CPUs such as Intel's Pentium continue to increase and costs continue to fall, we expect to see more and more features moved from voice cards to the host. Eventually, we may find that the only custom hardware needed is interfaces to the networks: IP network cards such as Ethernet for Voice-over-IP calls, and traditional trunk interfaces for the old-style telephone network.

5.11   Voice card programming

There are two main categories of programming interfaces (APIs) available for high-end voice cards: direct-to-board, also called native, and abstract.

Direct-to-board APIs are the lowest-level function calls provided by the voice card vendor, they are typically close to the firmware commands used to communicate between the host PC and the DSP software. Typical examples from Dialogic are the MNTI API for the DM/xx family of boards and the various families of functions for the previous generation of boards sometimes known as "R4" (functions with prefixes such as dx_-, dt_-, ms_- and so on).

Direct APIs may have less overhead and/or provide more control over low-level functionality of the board compared with abstract APIs. For example, a direct API might let you set a T1 robbed bit or send a user-to-user information element on ISDN at any time. An abstract API might hide such details behind higher-level functions such as "MakeCall", ruling out this level of customization. On the other hand, you may have to write a lot of board-specific code. For example, you might have to write three versions of your of call control code, one for analog, one for robbed-bit T1 and one for PRI ISDN. With an abstract API, these might all be supported by a single set of function calls and you might therefore only need to write your call control code once. Most direct APIs are delivered as libraries for the C programming language.

Abstract APIs provide higher-level functions. These are often easier to use and provide a higher degree of vendor- and/or technology-independence. Abstract APIs will often also provide client / server functionality so that the application can be on a separate PC from the voice cards, and/or share resources with other applications using the same server. Typical examples include TAPI for Windows, the ECTF S.100 API which is supported by Dialogic's CT Media product and JTAPI (Java Telephony API). Abstract APIs are available for C, but are more likely to provide object-oriented interfaces for languages such C++ and Java and/or language-neutral objects such as COM or CORBRA.

Programming to direct or abstract voice card APIs is an alternative to VoiceXML for telephony-enabling your Web site. A detailed discussion of these APIs is beyond the scope of this book.

5.12 Application categories

A PC with voice cards can be used to build a wide range of different systems. Some of the common categories are as follows.

5.12.1 Voice Mail

Sometimes called message store-and-forward, voice mail applications allow the caller to record a message for later retrieval by the mailbox owner.

5.12.2 Interactive Voice Response

Often abbreviated IVR, Interactive Voice Response is sometimes used to refer to any automated telephony system that responds with voice prompts and can accept some kind of user input. Usually, however, it is implied that there is some kind of database transaction involved. Banking by phone is the prototypical IVR application.

5.12.3 Audiotext

Provides a fixed "menu tree", usually touch-tone based, offering pre-recorded messages. Used for simple information systems.

5.12.4 Media Gateway

A media gateway is a converter between an IP network and a traditional telephone network. It will have network interfaces for both PSTN trunks and IP trunks, and may additionally have media processing resources for enhanced functionality such as touch-tone digit detection (touch-tones are lost when using highly compressed IP voice packets and must be transmitted out-of-band), PIN number collection and other automated user interactions.

5.12.5 Call Center

A common application of Computer Telephony Integration (controlling a switch by a separate computer) is found in a call center, e.g. a customer care center. Call centers have live operators which respond to in-bound calls, who for example accept orders from mail-order catalogues, or make out-bound calls such as opinion polling. A switch designed for a call center is often called an ACD (Automatic Call Distributor). ACDs offer specialized features such as queue management ("your call will be answered in the order received") which keeps overflow callers on hold and distributes them to agents as they become available. If the switch makes out-bound calls it may be called an auto-dialer, it will have algorithms to keep agents busy so that they don't waste time listening to call progress tones, repeated attempts to reach a number, and so on.

A call center may include an IVR system as well. This can be used to interact with a caller who is on hold waiting for an agent. The IVR system can offer status messages ("You are the seventh caller on hold, the average hold time is four minutes"), take messages ("To continue to hold, press 1, to leave a message, press 2"), offer information ("To hear more about our widgets, press 3") and so on.

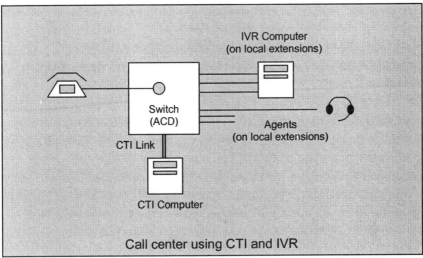

Call center using CTI and IVR

5.13   Anatomy of a VoiceXML browser platform

A typical VoiceXML browser is deployed on a PC or other industry-standard platform with an IP network interface card (NIC) and voice cards as trunk interfaces and/or media processors. A platform designed for a Voice over IP network will have one or more high-speed NICs playing the role of the trunk interface.

What we have been informally calling a "browser" can be broken down into logically separate modules called the interpreter and interpreter context.

5.13.1 VoiceXML interpreter

This is the software which reads and processes VoiceXML pages as described by the VoiceXML language standard. This piece is usually designed using abstract interfaces for interacting with telephony and IP network components. An abstract interface is a set of functions which the interpreter needs but which must be provided by another module. An abstract interface can be implemented as a C function library, Windows DLL, a C++ pure virtual class, a Java interface, COM or CORBA object and so on.

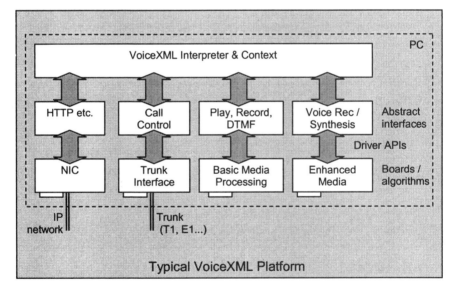

Typical functions in an abstract interface would be *PlayURL(URL)* to play a sound file, *ActivateGrammar(URL)* to activate a grammar, *DisconnectCall()* to hang up the call, and so on. There would also be reverse interfaces where an outside module calls into the interpreter (also called event or sink interfaces), typical functions here are *DTMFDetected(Digit)* to report when a touch-tone digit has been detected, and *GrammarMatched(GrammerInfo, Utterance,*

SlotValuePairs) to report that speech input matched an active grammar.

Details of the abstract interfaces are proprietary, they are designed by the browser manufacturer. In some cases, the interfaces may be published to allow third parties to do their own integration with new types of hardware or media processing algorithms, or they may be kept as internal details available only to the browser vendor.

5.13.2   Interpreter context

The principal responsibility of the interpreter context is handling the initialization and termination of an interpreter session.

A typical interpreter context accepts an incoming call on a T1 or E1 trunk and does a database lookup to determine the URL of the VoiceXML start page for the call. The URL could be determined by a combination of the ANI, DNIS, time of day, available system resources and other factors. Most common is a simple lookup based on the number the caller dialed, in other words the DNIS digits.

When the caller hangs up, the interpreter notifies the context. The typical response by the context will be to disconnect the call and make an entry in a billing or call record database.

6 Text-to-Speech

6.1   Speech synthesis

Speech synthesis, also called text-to-speech or TTS, is a technology
which generates artificial speech from text.

The following sections give a brief outline of the processing stages
found in a typical speech synthesizer. Details will vary
substantially between synthesizers from different vendors.

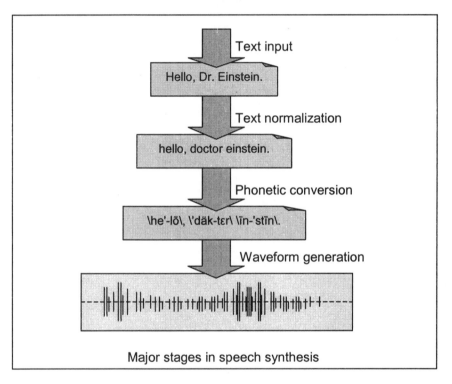

Major stages in speech synthesis

6.2   Text input

The text to be spoken is first read from a file or streaming buffer.
An input rate of something like 10 bytes per second is required to
give the average of about three spoken words per second or 180
words per minute which is typical for "read aloud" speech such as
a radio announcer.

During the text input stage the character set of the text may be
converted to the native format required by the synthesizer, which

for modern products is usually Unicode. Unicode is a 16-bit character set which covers most of the world's major written languages and has been widely adopted by operating system and application vendors.

6.3   Text normalization

The text normalization process expands abbreviations and resolves other ambiguous forms.

Abbreviations are often ambiguous: for example, should "Dr." be interpreted as "Doctor" (the title for an English-speaking physician or academic), "Dronning" (the Danish title for a Queen) or "Drive" (an address)?

Numbers in particular need special attention. "1994" should be spoken as "Nineteen ninety four" if it's a year, but "one nine nine four" if it's a taxicab number, "$1994" should be spoken as "nineteen hundred ninety four dollars."

The text "1/2" might be spoken "January second" (US), "February first" (UK), "one half" or "January two thousand two", again depending on the context in which it is used.

Consider this input text:

```
Dr. Einstein gave up 1/2 his income in taxes in
2001, paying $2001 to the IRS.
```

After normalization, the text might read:

```
doctor einstein gave up half his income in
taxes in two thousand one, paying two thousand
and one dollars to the [letter I] [letter R]
[letter S].
```

As these examples show, significant intelligence may be required in the generator to perform this step.

6.4   Exception dictionary lookup

Dictionaries are used to isolate exceptions to common pronunciation rules. The exact stage where the dictionary is used depends
on the detailed strategy used by the generator. For example, the
English word "of" is pronounced with a soft "v" rather than with
an "f" sound. The importance of the extent and quality of
exception dictionaries varies greatly with the language and
application. English has relatively many exceptions to spelling
pronunciation rules, especially when proper names in North
America must be pronounced: place names and peoples' names
are derived from many languages other than English. A language
like Norwegian, where spelling rules were defined much more
recently and which is therefore more consistent, may require less
attention from exception dictionaries.

6.5   Conversion to phonetic spelling

The speech synthesizer must arrive at a phonetic spelling for each
word in the phrase to be spoken. This can be done using rule-
based and look-up techniques. Look-up is conceptually simpler:
the generator has a table which converts conventional spelling to
phonetic spelling.

As an example of such a table that you might have at hand,
Webster's Collegiate Dictionary, provides phonetic spellings. The
speech synthesizer will use a similar representation internally to
describe pronunciation. For example, Webster's spells "alphabet"
phonetically as \ˈal-fɛ-bɛt\. Hyphens are used to separate
syllables. The quote mark (') indicates that the stress is at the
beginning of the word (one says "<u>al</u>phabet", not "al<u>pha</u>bet" or
"alpha<u>bet</u>"). A phonetic alphabet is used to describe in detail how
the word is spoken where the various vowel and consonant
sounds are assigned specific marks. The "s" in \ˈvóis\ for "voice",
for example, represents the hard "s" as in "kiss" rather than the soft
"s" as in "noise". There are various phonetic alphabets used by
linguists, one of the best known is the International Phonetic
Alphabet (IPA) which is included in the Unicode character set.
Remember that each variant of each word must be included: it is
not enough to store speak=\ˈspék\: the words speaks, spoke,

spoken, and speaking must also be translated. A phonetic dictionary must therefore be quite large to systematically cover a reasonable subset of a language.

By contrast, the rule-based approach attempts to derive the pronunciation from the conventional spelling. For example, it could be a rule that "c" would become phonetic "k" (as in car, record), unless followed by "ei", in which case it becomes phonetic "s" (ceiling, receive). In a language such as English with irregular spelling, a long list of rules and a long list of exceptions are needed for this approach to be effective.

In practice, a combination of the phonetic dictionary for some subset of the language combined with a rule-based approach for unknown words is likely to be used.

6.6   Phonetic modification

The phonetic spelling in a conventional phonetic dictionary may not fully specify the precise sound which would be made by a person. For example, the "t" sound in "cat" is subtly different from the "t" sound in "tom", but this is not usually distinguished by a phonetic spelling.

The phenomenon called co-articulation should also be considered: for example, a person will not speak two "d" sounds when saying "bad dog" or two "n"s in "seven nine".

These kinds of issue are typically resolved in a phonetic modification step.

6.7   Inflection

The overall shape of the phrase determining the pitch, weight (emphasis) and varying speaking rate of the voice is known as the inflection or prosody of the phrase. One factor determining the inflection will be whether the phrase is an exclamation, question or statement. Consider the following three sentences:

```
Hello, world.
Hello, world!
```

```
Hello, world?
```

These sentences have essentially the same phonetic components but should be spoken with different inflections.

6.8   Waveform generation

The final stage of synthesizing speech is to generate a waveform from an internal representation of the phonetic content and prosody. This may be done by concatenating together stored elements, by using algorithms which use parameter-driven rules, or a combination of the two.

Stored elements may include phonemes, which are the smallest units of identifiably different sounds roughly corresponding to letters such as "b" or pairs of letters spoken as one sound such as "th" or "ee", and disyllables, also called diphones, which are pairs of phonemes. Typically the database of phonemes and other sounds will have been carefully extracted from pre-recorded speech by a professional voice talent. There are about 44 phonemes in the English languages (the number varies according to how picky you are about distinguishing sounds and who is doing the analysis). Using diphones tends to improve the sound of the output waveform since the sound of a phoneme tends to change subtly as it transitions into the next sound: if two or more phonemes are stored as a single unit, this effect is automatically taken into account.

Rule-based generation starts from a model of the human vocal tract and runs what in effect is a simulation of the tongue, lip and throat actions involved in speaking.

6.9   Constructing phrases with variable values

In general, it is preferable to play pre-recorded prompts rather than using speech synthesis. The artificial voice can sound robotic and become tiring if listened to for long periods. Sometimes, as with e-mail readers, there is really no choice. If you have phrases which fit a fixed template needing variable values (for example, "your balance is two dollars and three cents"), you can use pre-recorded components, though there i more effort is needed to prepare the recordings and define the right algoriths to combine them. Some or all of this may be done internally for you by a good browser for some common data types, but even in those cases you might still need to prepare the correct vocabulary files so that the voice matches the rest of your application.

Some examples of such phrases which are constructed on the fly in telephony systems are as follows.

6.9.1 Mailbox status

"You have twenty-three new messages and five saved messages in your mailbox".

6.9.2 Date and time stamps

"This message was recorded at four thirty p.m. on July second."

6.9.3 Time, temperature and weather readings

"The time is now eight twenty seven a.m., the temperature is sixty-eight degrees, the wind is ten miles per hour from the East and the surf is light to moderate."

6.9.4 Account balance information

"Your current account balance is two thousand, five hundred and twenty three dollars. Your most recent cleared check was number one hundred two in the amount of thirty seven dollars and six cents."

6.9.5 Telephone numbers

"The number you are calling has been changed, the new number is 555-1234, please make a note of it".

6.10 Phrase components

Typically, the phrases which are constructed are composed of sentences with blanks where a few common elements are inserted. For example,

"You have ____ new messages."

Here, the blank is filled by a whole number (zero, one, two, ...).
The phrase is constructed from the partial sentence:

"You have...",

followed by the whole number, followed by,

"..new messages."

The recording "You have.." would be made with a slight upward
intonation so that the listener expects more to follow, and "..new
messages" would be recorded with an intonation indicating the
end of the sentence.

Fitting these pieces together can produce excellent results, but
even the most careful production will have a slightly artificial feel
to it. If there are only a limited number of possibilities, it may give
a more natural-sounding result to record all possibilities as
complete sentences. Suppose, for example, that your voice mail
system cannot store more than 99 messages. In that case, it might
be worth recording 100 complete sentences:

"You have no messages."
"You have one new message."
"You have two new messages."
...
"You have ninety-nine new messages."

The appropriate phrase to play would then be selected based on
the number of messages. Note that, as this example shows, there
are in fact two or more fragments which need to be recorded for
the end of the phrase: it should be "you have one message", not
"you have one messages". The complete phrase "you have no
messages" should probably be recorded as a special case.

Even better would be to select a different wording for the case of
an empty mailbox, for example: "this mailbox is empty." This

allows the experienced caller to get the information without having to listen through "you have..", he or she will recognize the beginning as soon as he or she hears "this mail..." and have the option of immediately selecting a different menu item or hanging up.

Commonly occurring phrase components are as follows.

Whole numbers
One, two, three ...

Ordinal numbers
First, second, third...

Dates
January 1st, or perhaps January 1st 1993.

Times
Eleven thirty five p.m.

Date/Time stamps
January 1st at eleven thirty five p.m.

Money
Sixteen dollars and fourteen cents.

Digit and alpha-numeric strings
For example, account numbers, where each digit is pronounced separately (so 1234 is pronounced "one two three four" rather than "one thousand two hundred thirty four").

Phone numbers
These are a special case of digit strings. Phone numbers often have fixed lengths (say, three for a PBX extension, seven for a local number and ten for a long-distance number).

There are special considerations which may apply to each type of phrase component.

6.11 Whole numbers

All whole numbers in English can be spoken with the following vocabulary.

```
Zero One Two .. Nine
Ten Eleven .. Ninteen
Twenty Thirty .. Ninety
Hundred
Thousand
Million
Billion
```

This represents a minimal set, a total of only 33 vocabulary files — sometimes it is useful to reduce the amount of memory needed, e.g. for a solid-sate appliance device.

By adding more files, the quality of the generated numbers may be improved. For example, each whole number from zero to 99 might be recorded as a separate file: a single recording of "twenty four" will sound better than the concatenated pair "twenty" "four". The numbers 1 .. 99 represent the most commonly used numbers in most applications and could be recorded as special cases. If not all 99, perhaps the first twenty or forty numbers.

Non-English languages can be approached in a similar way, but the details of the algorithm are likely to be different. For example, French speaks 21 as "vingt-et-un" (literally "twenty and one"), Danish speaks 21 as "to-og-tyve" (literally "two and twenty"). Each language may give different trade-offs in memory usage, quality of output and difficulty of preparing the vocabulary files.

Special cases which often need to be considered are zero and one. In some situations, it might be better to say "no" rather than "zero": for example, "you have no messages" rather than "you have zero messages". As mentioned earlier, the zero case often deserves its own recording, such as "this mailbox is empty". The case of one also changes the following noun from singular to plural: "one message" but "two messages".

In many non-English languages, there are two or more different varieties of "one" which depend on the noun. For example, Danish has both "et" and "en" (neuter and gendered), thus "et hus" (one house) uses a different variety than "en hund" (one dog). This can impact not only the algorithms needed to speak numbers but also database information: a database field may be needed to indicate whether a thing is masculine or feminine, neuter or gendered in the local language. Systems which cater to multi-lingual users must take several such complications into account.

6.12   Ordinal numbers

The ordinals are the whole numbers as used to indicate an ordering or ranking: "first", "second", "third" and so on.

Adding 32 new vocabulary files is all that is required to speak all ordinals.

```
First Second .. Ninth
Tenth Eleventh .. Ninteenth
Twentieth.. Ninetieth
Hundredth
Thousandth
Millionth
Billionth
```

Many applications will use ordinals only for dates, in which case it makes sense to record each ordinal from "first" to "thirty-first" as a separate file.

6.13   Dates

Speaking dates can quickly be accomplished by adding twelve more vocabulary files, one for each month.

```
January .. December
```

You might consider dropping the year or saying "this year" for dates in the same year. For example, if today is 8/8/2003, then the date 6/6/2003 would be spoken as "June sixth", but "5/5/2002" would be spoken as "March fifth two thousand two."

6.14   Times

Speaking times can again build on the components we have
already discussed. There are perhaps only four new vocabulary
files we need:

```
a.m.
p.m.
midnight
noon
```

As the above list suggests, the times 12:00 a.m. and 12:00 p.m.
might be considered special cases and spoken as "twelve
midnight" and "twelve noon" (this helps people like the author
who have trouble remembering if "twelve p.m." is noon or
midnight). Some non-English languages, such as Danish, use a 24-
hour clock in colloquial speech; for these languages no special
cases might be needed at all.

6.15   Date and time stamps

A date and time stamp is the moment when an event occurred,
such as the recording of a message, as stored on the computer.
The simplest approach is just to speak the date followed by the
time (or vice versa). However, it might be annoying to get such a
long phrase. For example, if I am listening to my voice mail
messages, chances are that most of them were recorded today. It
will significantly add to the time it takes me to listen to all my
messages if I must sit through "twelve thirty five p.m. August
eighteenth nineteen ninety three", or similar, before each message.
In this type of situation, it might be best to check for some special
cases. For example, when playing the date and time for a voice
mail message, one of the following approaches might be taken.

6.15.1   Today

Play the time only. Perhaps say "today" as a confirmation.

6.15.2   Yesterday or previous business day

Play the time followed by "yesterday" (much quicker than a date), or the name of the day such as "Friday", especially if a business holiday has intervened.

6.15.3   Within the last month

Play the time, day of month and month.

6.15.4   Older than the last month

Play the time, day of month, month and year.

If date/time stamps older than a month are going to be common, it is worth considering whether the time of day is important or whether it could be omitted, saving the time needed to speak a complete date. In a voice mail system, this is likely to be a rare exception, so a sensible default is to provide all the information.

6.16   Money amounts

Money amounts can again easily be spoken by building on the routines we've already discussed. We just need to add:

```
dollar
dollars
cent
cents
```

And perhaps:

```
no cents
exactly
```

Some might prefer to add the word "and" between the dollar and cent amounts.

Special cases to consider are to use the singular where there is $1 or 1¢ (use "Dollar" instead of "Dollars", "Cent" instead of "Cents"), and the following.

$0.00
This might be spoken as Zero Dollars Zero Cents or as Zero Dollars No Cents.

$0.CC
This might be spoken as *CC* Cent(s), or as Zero Dollars *CC* Cents.

$DDD.00
This might be spoken as *DDD* Dollar(s) exactly, or *DDD* Dollar(s) and no Cents.

6.17   Digit strings

To speak a long string such as a credit card number, consider inserting short pauses to group the digits. Groups of four digits might be a good choice for Visa and Mastercard numbers, for example.

English speakers use both "zero" and "oh" for the digit 0 in a string. A more formal system, such as banking by phone, might choose to speak "zero" in an account number since it sounds more official. An entertainment program might choose the more friendly sounding "oh": "your lucky number is two oh one!".

Where critical information must be provided, such as in an emergency notification system or applications to be used in noisy environments, unambiguous pronunciation techniques might be considered. An example is the military "niner" to distinguish "nine" from the similar sounding "five". With alphanumeric strings, the internationally standardized phonetic alphabet (Alpha, Beta, Charlie, Echo...Zulu) might be used for the letters, so that "P19" could be spoken "Papa One Niner". Special cases which sound similar, such as "thirteen" and "thirty", might also be a consideration when accuracy is important in a less than ideal listening environment.

6.18   Phone numbers

Phone numbers are special cases of digit strings. Again, it might be worth grouping the numbers according to the usual convention. A local number (NXX-YYYY) in the US might have a pause after the first three digits, a long-distance number (AAA-NXX-YYYY) might have pauses after the third and sixth digits.

Again, the choice must be made between "zero" or "oh" for speaking "0". In the US, "oh" is the usual choice, "601 1234" would be "six oh one...".

Phone numbers also provide opportunities for checking special cases as a person would. For example, the phone number "601 2233" might be spoken as "six oh one double two double three". This technique would not generally be applied to other situations, such as account numbers, which also speak digit strings.

Other languages have their own conventions for speaking telephone numbers. The English name each digit, but the Danes look at digit pairs so that the number "01 23 43.." would be spoken (literally translated) as "zero one, twenty three, forty three ..".

6.19   Inflection

If only one recording (vocabulary file) is used for each digit, month etc., then the resulting numbers and dates produced will have a monotone sound since the tone of voice and pitch will not vary as the phrase is spoken. Natural speech varies tone and pitch, giving the listener cues for positions in the phrase and the information being conveyed. For example, an English speaker will often raise the pitch of his or her voice at the end of a question, or lower the pitch of the voice at the end of a normal sentence. In contrast, a speaker in many Indian languages raises his or her voice at the end of a normal sentence. When speaking a phone number, an English speaker will speak the digits in groups with a slight upward intonation at the end of each group except the last, where the tone of voice will drop. These variations in the voice are called *inflection*. Phrases generated by voice response systems can add inflection as an ingredient to make the resulting speech sound more natural.

An inflected computer voice probably familiar to most readers is used by the directory information services (411 or 555-1212) in the US to speak the number back to the caller. Both the greeting ("Hello, my name is Fred, what city please?") and the telephone number found are machine-generated, minimizing the time spent by each operator on a call. The technique used by such a system is to record each digit, month etc. with several different inflections, the particular recording chosen for each digit depends on the digit's position within the phone number.

6.20 Preparing vocabulary files

Even when you've got all your software routines together, there is still the issue of recording and fine-tuning all your voice element files. You'll need to make sure they're trimmed (silence removed from the beginning and end), normalized (all have the same volume), and play-off filtered (touch-tone frequencies removed).

Remember too that your application will evolve with time: better sign your voice talent to a long-term contract, or you'll end up with several voices in your phrases. This common malady has been described as the voice mail ransom note.

```
Don't let THIS happen to you!
```

7 Speech Recognition

7.1 Voice recognition

Technology for voice recognition, (VR) otherwise known as
speech-to-text, speech recognition, automatic speech recognition
(ASR) or speaker independent voice recognition (SIVR)
technology has been improving rapidly. Just a few years ago, a
typical telephony voice recognition technology would probably
have the following characteristics.

7.1.1 Limited vocabulary

The maximum vocabulary size was typically in the range of ten to
perhaps 50 words.

7.1.2 Discrete recognition

The recognizer would typically require that the user pause
slightly between each word. There were a few exceptions: some
systems would allow continuously spoken digits, for example.

7.1.3 Pre-built vocabularies

The technology would typically require the system developer to
collect a large number of samples of the words in the desired
vocabulary. These samples would be analyzed to create a loadable
vocabulary file.

7.1.4 High error rates, low tolerance

Recognizers often had high error rates (mis-identified words,
words not recognized) and a low tolerance for regional accents
and other variations in speech (e.g., when the speaker has a cold),
especially when the speaker's accent was outside the range used
to create the vocabulary.

7.1.5 Limited language support

Many recognizers supported only English plus perhaps a few
major European languages, such as French, German and Spanish.

7.1.6 No voice interruption

Often the user would have to listen to an entire prompt before
speaking, there would be no ability to interrupt a prompt by

speaking a response early. This can be frustrating, especially for experienced users.

7.2 Modern recognizers

Today's recognizers have improved dramatically and offer capabilities which go far beyond those of a few years ago.

7.2.1 Large vocabularies

Recognizers are now capable of recognizing very large vocabularies. Accuracy is improved if the vocabularies are limited to a smaller sub-set relevant for a particular point in the application.

7.2.2 Continuous recognition

The user generally does not have to pause between words.

7.2.3 Grammars

A grammar is a template, typically using the standard written form of a human language, which specifies combinations of words to be match. A grammar is similar to a regular expression matcher in a text editor: it specifies a set of possible strings to match. A simple grammar might look like this.

```
$color = ( red | green | blue );
```

This would match if the user spoke "red", "green" or "blue". Grammar compilers figure out the pronunciation of the written words in the grammar, these are matched against the phonetic representation of the speech provided as input to the recognizer.

Grammars can now be compiled so quickly that they can be specified in real-time (or close to it) as an application progresses. This feature is emphasized in the design of VoiceXML.

For a few specialized and important purposes, such as highly reliable yes/no discrimination, grammars may still be pre-compiled and hand-tuned, but in most cases there is now little need to compile large sets of samples and build vocabularies before performing successful recognition.

7.2.4   Reduced error rates

As algorithms have improved and processing power has increased, error rates have continued to improve. On average, error rates have been dropping at about 30% per year since the early 1980s when computer-based speech recognition research began to reach commercial products.

7.2.5   More languages

Modern recognizers offer a much wider range of human languages.

7.2.6   Barge-in support

Current recognizers support barge-in, in other words the ability to interrupt a prompt by speaking a response early.

7.3   Voice recognition stages

A voice recognizer typically moves through the stages described in the following sections. As you would expect, the details vary among different technologies.

7.4   Capture, echo cancellation and endpointing

The first stage is to capture digital samples of the speaker's voice. In a telephony system, the digitization is done for you either by the telephone network or by an analog trunk interface. It is much better to use a digital interface, meaning that you should use a T1 or E1 (or perhaps BRI) trunk. If you use an analog interface, then the voice quality may be significantly degraded because the sounds will probably have gone through at least three conversion stages: analog to digital at the caller's central office switch, digital to analog at your local switch, and then analog back to digital on your trunk interface card. Each conversion step is likely to degrade the signal. It is important to get the best possible voice quality to achieve good recognition, so using digital trunks is almost an essential for a commercial system.

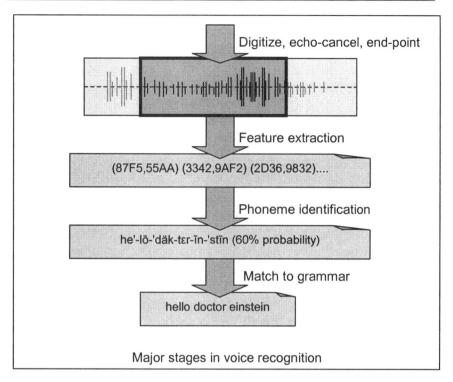

Digitize, echo-cancel, end-point

Feature extraction

(87F5,55AA) (3342,9AF2) (2D36,9832)....

Phoneme identification

he'-lō-'däk-tɛr-īn-'stīn (60% probability)

Match to grammar

hello doctor einstein

Major stages in voice recognition

This stage also performs two other important functions: echo cancellation, which removes any echoes of the computer's prompt, and end-pointing, which identifies the start and end of the utterance to be recognized. We will have more to say about these functions later.

7.5   Feature extraction

The echo-canceled, end-pointed speech is typically represented in a format such as 8-bit, 8 kHz linear PCM. The feature extraction stage analyzes the audio to create a numerical representation which more closely corresponds to sounds which are perceived by the human ear. For example, a typical operation performed is Fourier analysis, which analyzes the frequency components of the sound.

The output from this stage is usually a so-called spectral analysis, a set of numbers which represents the pitch combinations and how they vary with time.

7.6 Phoneme modeling

The phoneme modeling stage attempts to convert the spectral analysis into possible sets of phonemes. This is the most difficult phase in voice recognition technologies and the quality of the phoneme modeling algorithms is often the key to determining the relative success of one recognizer against another.

The output from this stage is a set of candidate recognitions expressed as strings of phonemes, perhaps with additional information which can indicate segmentation, i.e. possible boundaries between syllables and words, stress, inflection (prosody) and other useful indicators. Each candidate recognition will have a weight or score, a measure of probability that the recognizer has assigned to whether this candidate is a close match to the speech.

7.7 Grammar matching

The final stage matches the candidate strings of phonemes against one or more active grammars. Each grammar in effect represents a regular expression of phonemes which is matched against the candidates in a similar fashion to the advanced search feature in a programmer text editor.

The output from the final stage will be one or more proposed utterances, probably expressed as legal strings produced by one of the grammars. For example, imagine the phoneme modeling stage produced two candidate phoneme strings, "red" and "rɛd":

```
red       Weight 60%
rɛd       Weight 30%
(Other)   Weight 10%
```

Now imagine these phonemes were matched against this grammar:

```
$color = ( red | green | blue );
```

The output from the final stage might be:

```
red      Weight 95%
```

7.8   Telephony issues for voice recognition

Voice recognition over the telephone is significantly harder than recognition based on a desktop microphone. The issues include limited bandwidth, microphone quality, and echo cancellation.

7.8.1   Limited bandwidth

The public telephone network carries digital audio encoded as 8-bit samples at a rate of 8 kHz (8,000 samples per second) for a data rate of 64,000 bits per second. By contrast, a music CD has a data rate of 705,600 samples per second for each stereo channel, more than ten times more.

While the network carries digital sound with little or no loss in quality, the "last mile" to the subscriber's home or office is often based on older, analog technology which can further degrade the quality.

A well-known result in digital signal processing, Nyquist's Theorem, says that the maximum frequency carried by a channel cannot be more than half the sampling rate. This tells you that even an ideal, all-digital phone connection cannot carry audio frequencies above 4 kHz. Unfortunately, the human ear is sensitive up to about 20 kHz, and human speech does make use of frequencies above 4 kHz to distinguish some sounds, such as "S" and "F".

The end result is that the audio delivered over the public telephone network is far from ideal for automated voice recognition—even those highly advanced recognizers called people often have problems in distinguishing similar words over the phone.

7.8.2   Microphone quality

Most telephones have poor quality, inexpensive microphones. This compounds the problem caused by limited network bandwidth.

7.8.3   Echo cancellation

In desktop voice recognition, output is usually displayed on a screen—it uses a separate "channel" and therefore does not interfere with the user's spoken input. With an automated telephone system, output from the computer also uses an audio channel. If there is an analog leg anywhere in the path between the caller and the voice recognizer (which is almost always the case since most callers will have analog phones), then the recognizer must deal with echo. If the computer is playing a prompt, then the sound of that prompt "loops around" the analog connection and is also heard by the recognizer, this is called echo. The sound of the prompt will be mixed with the caller's voice if he or she is speaking. The recognizer must subtract the prompt currently being played in order to isolate the caller's voice before beginning recognition, this subtraction process is called echo cancellation. Unfortunately, echo cancellation is difficult because the echoed sound will be distorted and slightly delayed by its trip around the network, so the subtraction will inevitably be less than perfect.

7.9   User interface design

The design of user interfaces for voice systems and graphical systems have many parallels but also key differences. Voice interfaces are in many ways more challenging.

One key problem is the extremely limited bandwidth of the connection, which means that at most only a few bytes per second of meaningful data can be transmitted per second in each direction.

Also, the interface must rely on user's short-term memory to retain the recent history of the conversation, for example the list of choices which the computer has just spoken in a menu.

By contrast, a graphical interface can present thousands of bytes of meaningful data in such a way it can be scanned by the user. The presentation can show the user many commands within a short time without relying on memory, as when a user selects an option from a pull-down menu.

Designers recognize two major categories of interface: user-directed and computer- or machine-directed. In a computer-directed interface, the computer asks questions or demands input and the user responds to these requests. A typical example might be a login prompt, where the computer says "type Ctrl+Alt+Del to log in", the user responds by typing that key combination. The computer then presents a dialog or form which requires the user to type in a name and password. In a user-directed interface, the roles are reversed: the user issues requests and the computer responds. A typical word processor application is user-directed most of the time: the user can for example pull down menus or type keystrokes such as Ctrl+I to italicize text, the computer responds to those commands.

The application may alternate between those modes, in which case it is said to have a mixed-initiative interface. A typical example in a word processing application happens when the user issues a (user-directed) *Quit* command without saving the most recent version of the document. The computer will respond with a (machine-directed) dialog requiring the user to select one from a list of choices such as *Save Before Quiting, Keep Editing* or *Quit Without Saving.*

Historically, automated telephony systems have been mostly machine-directed: the user generally has to respond to the computer's requests to dial a menu selection or an account number, with perhaps minor exceptions such as the ability to dial the star digit "*" at any time to return to the main menu.

VoiceXML is designed to allow all three types of interface: user-directed, machine-directed and mixed-initiative, as we shall see.

This should help application developers created new classes of more user-friendly telephony applications.

Designing good telephony interfaces will remain a challenging problem, but one we will not address in any depth in this book. Our goal here is to provide a clear explanation of the VoiceXML language; our example code will emphasize ease of understanding and will in most cases not be suitable to be applied directly in a high-quality application. Explaining good user interface design is another goal for another book.

7.10 Dialog components

Variously called Dialog Components (DCs), Dialog Application Components (DACs), Speech Objects (the term used by Nuance Communications) or Dialog Modules (used by Speechworks International), these are pre-compiled voice dialog algorithms.

To introduce the idea, we'll describe a simplified example. The upcoming flowchart shows an algorithm for getting a Yes/No answer. "Q" is the question to be asked: for example, it might be "Are you sure you want to close your account?". Note that this algorithm will say "Did you say Yes?" in the case where "No" was matched, but with low confidence. This is less likely to confuse the user than asking "Did you say No?" or repeating the question. Following "Did you say Yes?", the algorithm accepts Yes or No even with low confidence, providing it matches the first match (for simplicity, the above flow-chart does not handle the case where there is no input or no match to this question). Since this is not a book about good user interface design, we'll simply accept that this is a good algorithm.

The idea is that algorithms for collecting common types of values, such as Yes/No, date, time, credit card number and so on can be developed by experts in voice user interface design and then re-used in different applications. In this way, the application designer doesn't have to be an expert in voice user interface design and doesn't have to spend time re-inventing common user interface tasks.

DCs are algorithms like this flow-chart. They are usually written in traditional programming languages like C, C++ or Java. They are compiled into independent modules such as S.100 applets, Java class files, Java Beans, or Windows DLLs, or are created in industry standard object frameworks such as COM, ActiveX or CORBA.

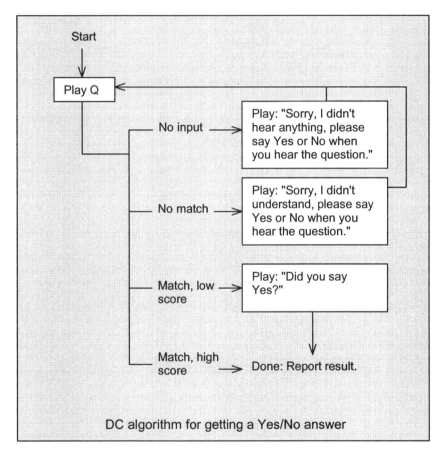

DC algorithm for getting a Yes/No answer

DCs will typically need one or more input parameters and provide at least one output parameter. In this simple example the name of the prompt file Q to ask the question would be provided as a parameter. Perhaps the prompt file names for the error messages ("Sorry, ...") would also be passed in as parameters. The output parameter would be the result: Yes, No or Error (unable to get a reliable answer).

A well-designed DC architecture provides abstract, replaceable interfaces for input/output, i.e. for playing prompts and collecting touch-tone digit and voice recognition results. This allows the maximum flexibility in re-using dialog algorithms on different platforms.

VoiceXML is designed as a language for creating dialogs and re-usable dialog components. A browser vendor can choose to support DCs created using other architectures via the *<object>* tag.

8 VoiceXML: The Basics

8.1   Hello, Web

Ever since the appearance of Brian Kernighan and Dennis
Ritchie's classic book *The C Programming Language*, it has been
almost mandatory to introduce a new language by giving a Hello,
World sample. So here it is: a VoiceXML page which speaks
"Hello, World" and hangs up.

```
<?xml version="1.0"?>
<vxml version="2.0">
    <form>
        <block>
            <prompt>
                Hello, World.
            </prompt>
            <disconnect/>
        </block>
    </form>
</vxml>
```

In VoiceXML, the terms "page" and "document" are synonymous.
In this case, the page is also a complete application. An application will usually include other pages which are brought in via
hyperlinks, as we will see later. We'll assume that this page is
stored on the Web server in a file named *HelloWorld.vxml*.

At the core of this page is the *prompt* tag, which instructs the
browser to speak the text value of the tag (that is, the text between
the start and end tags) using text-to-speech.

As you might guess, *<disconnect>* ends the browser session and
hangs up the call.

(Typographical note: we often use *<tagname>* in descriptive text as
a synonym for "the *tagname* tag". This should not necessarily be
understood to indicate the correct syntax, which in this case is
<disconnect/>).

The rest of the code is scaffolding we need so that we have a place to put the *prompt* and *disconnect* tags. First is a standard XML declaration:

```
<?xml version="1.0"?>
```

This is required for any XML document; a VoiceXML page is no exception.

Next is the *vxml* tag. Every VoiceXML page is enclosed in one of these tags. For this reason, *<vxml>* is called the root tag for the document. The *version* attribute of *<vxml>* is required, we use "2.0" since VoiceXML 2.0 is the version described in this book (see Preface for more about this point).

For now, we are not going to discuss *<form>* and *<block>*, it is enough to know that code inside *<form><block>...</block></form>* is executed when this page is loaded. This gives us a framework in which we can try out what in VoiceXML jargon is called "executable content". That means tags which perform step-by-step instructions in the same way as traditional procedural code.

Typical examples of tags which can be used in executable content are the *prompt* tag, which is an instruction to speak its content (the text between the start and end tag) using text-to-speech, and *<disconnect>*, which hangs up the call and terminates the browser session.

8.2   Comments

Comments can be inserted using the standard XML syntax.

```
<-- This is a VoiceXML comment -->
```

8.3   Prompting with text-to-speech

As we have just seen, the *prompt* tag is a command to speak text using speech synthesis, also known as text-to-speech.

The text to be spoken can be "marked up" by tags which request particular speaking styles. These are analogous to mark-up tags in

HTML which specify font size, bold, italic and other attributes of text for a graphical browser.

For example, *<emphasis>* says that a given segment of text should be spoken with emphasis, this can be used like italics or bold in displayed text.

```
<prompt>
    I <emphasis>must</emphasis> learn VoiceXML.
</prompt>
```

The *break* tag inserts a pause. You can specify the length of time by using the *size* attribute, which takes the value *"large"*, *"medium"*, *"small"* or *"none"*, or the *time* attribute, which takes a numerical value followed by "s" for seconds or "ms" for milliseconds. The default is *size="medium"*.

```
<prompt>
    Let me think for a moment.
    <break time="1.5s"/>I'll choose tails.
</prompt>
```

In most places where you can put a *prompt* tag, you can simply write text, the browser understands that this text is to be spoken. In our first example, we used:

```
<block>
    <prompt>
        Hello, World.
    </prompt>
    <disconnect/>
</block>
```

Exactly equivalent is:

```
<block>
    Hello, World.
    <disconnect/>
</block>
```

However, when you use this short cut and omit the *<prompt>* tag, you cannot use mark-up on the text. So this is legal:

```
<block>
   <prompt>
      Hello, <emphasis>World!</emphasis>
   </prompt>
</block>
```

But this is not legal:

```
<-- ERROR! Don't do this! -->
<block>
   Hello, <emphasis>World!</emphasis>
</block>
```

A coding style which always uses *<prompt>*, even when it would be legal to drop it, is more robust against changes, and this is the style we use in our examples.

To get a complete overview of what you can write inside a *prompt* tag, refer to the Content Syntax section for this tag in the Tags Reference.

8.4   Prompting with sound files

To play pre-recorded sound, the *audio* tag is used. The file to be played is specified as the *src* attribute. (You might think of this as being similar to the *img* tag in HTML, where the *src* attribute specifies the URL of the image).

If we had recorded "Hello, World" into a sound file named *HelloWorld.wav* then we could re-write our Hello, World sample like this.

```
<?xml version="1.0"?>
<vxml version="2.0">
   <form>
      <block>
         <audio src="HelloWorld.wav"/>
         <disconnect/>
      </block>
   </form>
</vxml>
```

The value of the *src* attribute is the URL of the sound file. (More accurately, the URI, but we will use the more familiar term URL, which for most practical purposes is the same thing with today's Web servers).

The URL we specified has no path component, so it is interpreted relative to the URL of the enclosing page. The sound file could also be in a different directory or on a different server, as in the following example, which plays two consecutive sound files from different locations.

```
<block>
   <audio src="../SoundFiles/Hello.wav"/>
   <audio src=
     "http://www.demo.com/files/World.wav"/>
</block>
```

You can specify a text-to-speech prompt to be played if the sound file at the given URL is unavailable — the server is down or too slow to respond, for example. This feature is analogous to the *alt* attribute of the *img* tag in HTML, which specifies text to be displayed as an alternative to the image (while the image is downloading, as a mouse hover help display, or if the user has enabled a text-only option in the browser). This alternative prompt is specified as a value for the *audio* tag. If the sound file cannot be fetched, then the value will be processed exactly as if it had been in a *prompt* tag. Anything that is legal inside *<prompt>* is also legal inside *<audio>* and will be interpreted in the same way should the alternative prompt be required.

The following will speak "Hello, World" using text-to-speech if the sound file *HelloWorld.wav* is unavailable.

```
<block>
   <audio src="HelloWorld.wav">
      Hello, World.
   </audio>
<block/>
```

Some types of audio file, including Wave, have file headers which specify the encoding (bits per sample, samples per second and so on). Other types, such as Dialogic's Vox file type, contain just raw audio data. These are called raw files or headerless files. With such files, the browser must be told the encoding somehow. This is done in the HTTP response header which is sent by the server to the browser immediately before the audio data itself. The *Content-Type* field in the HTTP header must be set to the appropriate Media Type (sometimes called MIME Type), such as *audio/basic* (headerless 8-bit 8 kHz mu-law PCM). It is therefore the Web server which must know the encoding used in the file and report this to the browser, not the VoiceXML page. Most Web servers provide a simple mechanism for mapping file name extensions to Media Types so this is easy to implement—providing that you administer the server. If you are using a hosting service, this facility may be unavailable to you (because the file types used in telephony often do not have standard extensions); in which case you may find that using a file type with a header is a better solution.

8.5   Prompting with streaming audio

Streaming audio is sound which is produced in real-time, as opposed to being pre-recorded. A typical example is live commentary from a sporting event. The browser starts fetching audio data from the Web server when the play begins, and continues fetching until interrupted by the user or the Web server signals the end of the file.

A VoiceXML browser is usually permitted (indeed, encouraged) to keep local copies of resources such as sound files in a store known as a cache in order to reduce delays in responding to the user and reduce the load on the Web server. All tags which specify resources to be fetched have a common set of optional cache-related attributes, including *caching* and *fetchhint*. Audio is streamed by disabling the cache and instructing the browser to start playing without waiting for the complete resource to be downloaded, which is done by setting *caching="safe"* and *fetchhint="stream"*. Browsers are not obligated to support caching of any particular resource, but they are obligated to support streaming when caching is disabled, as in the following example.

```
<block>
   <audio src="http://realtime/livesports.raw"
      caching="safe" fetchhint="stream"/>
</block>
```

The default value for *caching* is *"fast"*, which suggests, but does not require, local caching.

8.6   Hyperlinking

A hyperlink is an element of a VoiceXML page which reacts to user input by transitioning to a new page or a new position within the current page. Hyperlinks are defined using the *link* tag. Two things are needed to define a hyperlink: 1) the user input which triggers the link, and 2) the target of the link.

User input is described using a *grammar* tag. A grammar is a template which defines a set of utterances which the user might speak. Following is a simple grammar which matches the word "sports".

```
<grammar>
   sports
</grammar>
```

One or more links can be specified as children of the *vxml* tag. In this case, the link is said to have document scope, and the link grammar will be "active" (i.e., monitoring incoming audio and prepared to match what the user is saying) unless something is done to disable it. Grammars can also be associated with other VoiceXML elements, as we shall see.

The target of the link is specified through the *next* attribute of the *link* tag, which should be set to the target's URL. Here is a hyperlink definition.

```
<link next="Sports.vxml">
   <grammar>
      sports
   </grammar>
</link>
```

This instructs the browser to jump to the page *Sports.vxml* if the word "sports" is spoken by the user.

Following is a complete VoiceXML page using this hyperlink. If the user says "music" at any time while this page is loaded, the browser will stop what it is doing and load the page *Music.vxml*.

```
<?xml version="1.0"?>
<vxml version="2.0">
   <link next="Sports.vxml">
      <grammar>
         sports
      </grammar>
   </link>
   <form>
      <block>
         <audio src="LongMessage.wav"/>
         <disconnect/>
      </block>
   </form>
</vxml>
```

A hyperlink may have more than one grammar. For another example, here is a hyperlink which accepts either the spoken word "sports" or the touch-tone digit 2. The *mode* attribute of the *grammar* tag is set to "*dtmf*" to indicate that it should match touch-tones, or to "*speech*" (the default) to indicate that it should match spoken words.

```
<link next="Sports.vxml">
   <grammar mode="dtmf">
      2
   </grammar>
   <grammar mode="speech">
      sports
   </grammar>
</link>
```

Grammars may be specified in two ways: in-line and externally. An in-line grammar is included in the page where it is used, as we have shown in the examples so far. An external grammar is referenced by using the *src* attribute of the *grammar* tag, which specifies the URL of the file where the grammar is stored. There is one grammar format which must be supported by a compliant VoiceXML 2.0 browser, it is called the Speech Recognition Grammar Format (SRGF) and is described in a separate chapter. SRGF is another XML language. SRGF grammars cannot be specified in-line; they must be in their own separate files. (This issue is under study at the time of writing and may change before the VoiceXML 2.0 standard is finally approved). There is an alternative form of SRGF called the Augmented Backus Naur Form (ABNF) which is optionally supported by VoiceXML browsers and may be specified in-line. Since it is easy to read, we will adopt the ABNF form in most of our examples.

8.7   Dialogs, forms and menus

Hyperlinks are a simple but limited way to collect input from the user. VoiceXML provides a much more powerful and flexible structure for managing user input called a dialog. There are two types of dialog: forms and menus. Menus add no new capabilities over forms, they provide a convenient syntactical shorthand for the special case where you want to provide a choice to the user.

As an example, we will construct a form which collects a credit card number and expiration date, then submits these values to a server script.

To do this, we need three items in the form:

- a field to collect the credit card number,
- a field to collect the expiration date, and
- a "submit button" which sends the information to the server.

Following is a VoiceXML page containing such a form.

```
<?xml version="1.0"?>
<vxml version="2.0">
   <form>
      <-- Credit Card Number field -->
      <field name="ccnr" type="number">
         <prompt>
            Please say or dial your credit
            card number.
         </prompt>
      </field>

      <-- Expiration Date field -->
      <field name="exp" type="date">
         <prompt>
            Please say or dial the
            expiration date.
         </prompt>
      </field>

      <-- "Submit button" -->
      <block>
         <submit next="ValidateCard.jsp"/>
      </block>
   </form>
</vxml>
```

The definition of a form is enclosed within a *form* tag.

The definition of a field which collects user input is enclosed within a *field* tag. The name of the field is specified by the *name* attribute. This is the name used in the *name=value&name=value&...* string in the URL (or form data) which is sent to the server when the form is submitted. The *type* attribute of the field specifies the data type of the value. Here we specified *type="number"* for the credit card, which means to expect individual digits like "one two three...", and *type="date"* for the expiration date.

A field is interpreted by first executing its content, which for both fields in this example means speaking a text-to-speech prompt, and then waiting for appropriate user input.

The *block* tag defines a "hidden" item, in other words, an item which does not correspond to user input. It is used to contain executable content. In this case, the executable content is a *submit* tag. By default, the submit tag sends all field values in the parent form to the server in exactly the same way as a Submit button (that is, a button defined using *<input type="submit">* in HTML). The server must respond with a new page, so *submit* is in effect a hyperlink.

The browser moves through the items in the order they appear in the page, so it first collects the credit card number, then the expiration date, then executes the block containing *<submit>*.

Items such as fields and blocks are variously referred to as items, dialog items or form items, all these terms mean essentially the same thing.

The conversation between the caller and the browser when interpreting this form might go as follows.

BROWSER: Please say or dial your credit card number.

CALLER *(dialing)*:   012345678901234

BROWSER: Please say or dial the expiration date.

CALLER *(speaking):*   Next month.

BROWSER: I'm sorry, I didn't understand that. Please
                        say something like "April seventh two
                        thousand seven".

CALLER: June two thousand three.

Notice that the browser was not able to understand the user input
"Next month" for the expiration date. In response, the browser
played a built-in error message, in this case one tailored
specifically for the date type. Only when a field has been
successfully filled in will the browser move on to the next field.

The *submit* tag has a *next* attribute, which specifies the URL. The
method attribute can be set to "*get*" (the default) or "*post*" to specify
which HTTP method to use. Using *POST* is recommended as good
style since the amount of data which can be sent using *GET* is
limited to the maximum length of the URL supported by the
server, but *POST* has no such limitation. You can submit selected
fields by using the *namelist* attribute which takes a value which is
a whitespace-separated list of field names. By default, all fields in
the form will be sent. So the following is exactly equivalent to the
submit tag in our example:

```
<submit next="ValidateCard.jsp" method="get"
  namelist="ccnr exp"/>
```

Based on the above conversation, this would generate a *GET*
command for the following URL:

```
ValidateCard.jsp?ccnr=012345678901234&exp=200306??
```

As this example shows, a date is formatted as an eight-digit string
"*YYYYMMDD*", fields which are not known are filled with "?"s.

8.8   Hyperlinks inside a dialog

A dialog may contain hyperlinks in addition to items by using *link*
tags. A hyperlink within a form is, by default, only active while

that form is being interpreted. This can be changed by the *scope="document"* attribute of the *link* tag, which makes it active for the entire time that the current page is loaded.

The following form has a hyperlink which reacts to the word "operator".

```
<form>
   <link next="Operator.vxml">
      <grammar>
         operator
      </grammar>
   </link>

   <field name="phonenr" type="digits">
      <prompt>
         Please say the phone number you
         wish to dial.
      </prompt>
   </field>

   <block>
      <submit next="DialNumber.jsp"/>
   </block>
</form>
```

The *goto* tag can also be used to hyperlink. This is similar to *<submit>*, except that no field values are transmitted. It can be used to jump to another form or menu in the same page by using the anchor notation *"#name"* familiar from HTML. (An anchor alone is never used with *<submit>* because the HTTP protocol requires that a new page be sent). If a dialog (form or menu) is to be the target of a link, it must be named using the *name* attribute, for example:

```
<form name="GetCreditCard">
```

The dialog can then be referenced by using this name as an anchor:

```
<goto next="#GetCreditCard">
```

8.9   Structure of a VoiceXML page

You will now be able to recognize that our first examples were based on a very simple form which had just one item, a block.

When a VoiceXML page is loaded, it begins executing the first form on that page (unless an anchor was used). In the case of our early examples, the one form on the page begins by executing its first (and only) item, the block.

The following picture may help you to visualize the structure of a typical VoiceXML page.

A page is composed of links and dialogs. The order in which they appear is not significant, with one exception: the first dialog to appear is the default start dialog.

A dialog is composed of links and items. The order in which the items appear is the default order in which they will be collected. We have so far introduced two types of item: *<field>* and *<block>*. Other items include *<record>*, *<transfer>*, *<object>*, *<subdialog>* and *<initial>*, which we will discuss later.

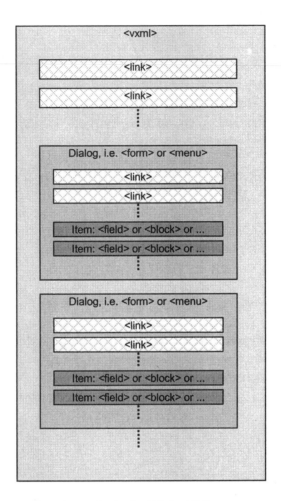

Structure of a typical VoiceXML page

This is not the complete story, but you may find it helpful to keep the picture in mind.

A dialog will generally be designed to hyperlink to another dialog by using *<submit>* or *<goto>*. If a form is completed but no hyperlink was triggered, the browser session ends. What happens when the session ends is up to the browser. Typically it will hang up the call.

If you want to force a hang-up, use *<disconnect>*. If you want to end the browser session and leave it up to the browser what to do next, use *<exit>*.

8.10 Built-in field types

We've already seen that a VoiceXML browser supports the built-in field types *number* and *date*. These types have built-in grammars for collecting input in the correct format and also have built-in error handling for re-prompting and coaching the user. The grammars and error-handling algorithms are browser-dependent; they are not specified by the VoiceXML standard. The following table is a complete list of the built-in types.

Type	Description
boolean	A yes/no or true/false value.
date	A date, including year, month and day or subsets such as year and month or month and day.
digits	Digit string, for example "one two three".
currency	Money amount.
number	Numerical value such as "one hundred twenty three".
phone	Phone number with optional extension.
time	Time of day.

The *digits* and *boolean* types support optional parameters, which
are specified using the following syntax.

```
typename?parm=value_{opt}; ?parm=value_{opt} ...
```

For example,

```
<field name="nr" type="boolean?y=0;n=1">
```

The optional parameters come after the type name following a
question mark "?". If there is more than one parameter, they are
separated by semi-colons ";". The parameters change the way the
input is collected.

With the exception of *boolean* fields, the value in the field variable
is used when performing a *<submit>*. With *boolean* fields, the
variable is stored as one of the strings *"yes"* or *"no"*, however
<submit> sends *true* or *false*.

8.10.1 Boolean

The *boolean* type accepts spoken phrases meaning yes/no
according to the locale. The locale is specified using the *xml:lang*
attribute, which will be discussed in more detail later. For
example, the browser might accept "yes", "yeah", "yep", "O.K.",
"no" and "nope".

By default, touch-tone input accepts 1 for yes and 2 for no. This
can be overridden by specifying the parameters *y=d* and/or *n=d*,
where *d* is a touch-tone digit. For example, the following field
accepts 0 for yes and 1 for no.

```
<field name="nr" type="boolean?y=0;n=1">
```

The field variable is set to a string value "yes" or "no", regardless
of the locale and input mode (speech or touch-tone). However,
when the value appears in a *<submit>*, it is sent using *true* or *false*,
so the above field might be sent as *nr=true*.

8.10.2 Date

The browser should accept spoken phrases that specify the following combinations:

```
year, month, day
year, month
month, day
```

Touch-tone input requires 8 digits for *YYYYMMDD*. The current VoiceXML draft standard does not permit touch-tone input for just year and month or month and day.

The field variable is set to an 8-digit string, also formatted as *YYYYMMDD*. If the year or day is omitted, it is specified as question marks "?", for example "????1225" or "200706??".

8.10.3 Digits

The *digit* type accepts strings of digits, either dialed as touch-tones or spoken as "one two three...".

There are three parameters: *minlength* (at least this number of digits are required), *maxlength* (at most this number of digits are required) and *length* (exactly this number of digits are required). For example,

```
<field name="credit_card_nr"
  type="digits?minlength=14;maxlength=16/>

<field name="US_zip_code" type="digits?length=5"/>
```

8.10.4 Currency

The *currency* type accepts a money amount. This may optionally include the currency unit, for example "one thousand francs". With touch-tone input, star "*" is accepted as the decimal point. The field value is set to a string in the format "*UUUhh.ff*" where *UUU* is the ISO 4217 symbol for the currency, *hh* is the whole number part (e.g. dollars) and *ff* is the fractional part (e.g. cents). Note that not all currencies have fractions based on one hundredth of the unit. Example ISO 4217 codes are *USD* for the

US dollar, *GBP* for the British point and *DKK* for the Danish crown. You can easily find the complete table by searching the Web.

8.10.5 Number

The *number* type accepts a spoken number such as "one hundred twenty three". Touch-tone input accepts digits and star "*" for a decimal point. Spoken input can specify a sign (e.g., "minus one"); there is no fixed convention for specifying negative numbers using touch-tones.

8.10.6 Phone

The *phone* type accepts a telephone number. When using touch-tone input, star "*" is understood to mean "extension". An extension is represented using the character *x* in the field variable. For example, the spoken phrase "four one five triple five twelve twelve extension one hundred twenty three" would be stored as "*4155551212x123*".

8.10.7 Time

The *time* type accepts a time-of-day in hours and minutes (not seconds). The result is a five-character string in the form *HHMMx*, where *x* is *a* for a.m., *p* for p.m., *h* to indicate that the 24-hour clock time was used or *?* to indicate that the input was ambiguous with regards to a.m. and p.m.. Touch-tone input is supported, however the standard does not specify a convention for collecting a.m./p.m., so a touch-tone result can be expected to end in *h* or *?*.

8.11 Customizing fields

VoiceXML allows you to create your own field types by specifying valid input, error handling, input validation and so on. In the following sections we will examine some of these features.

8.12 Field grammars

You can define a grammar for a field in exactly the same way as for a hyperlink. The grammar defines the acceptable utterances which may be spoken by the user as input for the field. Following is a grammar which accepts "red", "green" or "blue".

```
<grammar>
   red | green | blue
</grammar>
```

The vertical bar indicates a choice where exactly one of the
enclosed items must be present. A field named *color* using this
grammar could look like the following.

```
<field name="color">
   <grammar>
      red | green | blue
   </grammar>
   <prompt>
      Please say red, green or blue.
   </prompt>
</field>
```

The field value will be set to *red, green* or *blue* depending on what
the user spoke.

Grammars can be stored in separated files and referenced by a
URL using the *src* attribute of the *grammar* tag, as follows.

```
<field name="color">
   <grammar src="rgb.gram">
   <prompt>
      Please say red, green or blue.
   </prompt>
<field>
```

If the grammar is written using SRGF, the only grammar format
which a compliant browser must support, then according to the
current VoiceXML 2.0 draft, the grammar must be specified in a
separate file. (The author believes this restriction is likely to be
lifted in the final publication). The in-line examples we are giving
use the ABNF grammar format, which may or may not be
supported by a compliant browser.

8.13 Invalid user input

VoiceXML recognizes two main categories of incorrect input: "no input", meaning that neither the speech recognizer nor the touch-tone recognizer detected any input, and "no match", meaning that input was detected but did not match any active grammar with sufficiently high confidence. These are two examples of events, conditions which are reported by the browser and which can be "caught" and handled by the VoiceXML page. These events are named *noinput* and *nomatch*.

The *noinput* event is "triggered" or "raised" when a time-out occurs while waiting for spoken input following a prompt. The default time-out is not specified by VoiceXML; the length of time to wait is browser-dependent. You can specify an explicit value by using the *timeout* attribute of the *prompt* tag, as follows.

```
<prompt timeout="3.5s">
    Please say red, green or blue.
</prompt>
```

The value of *timeout* is a numerical value followed by "s" for seconds or "ms" for milliseconds.

The *audio* tag does not have a *timeout* attribute. To specify a time-out when using a pre-recorded file, enclose the *audio* tag in a *prompt* tag like this.

```
<prompt timeout="3.5s">
    <audio src="rgb.wav"/>
</prompt>
```

The *nomatch* event may be triggered because the user said something which really didn't match the grammar, or because the recognizer failed to correctly identify the input due to the user's accent, background noise or other factors.

These events are handled by placing executable code in a *nomatch* or *noinput* tag inside the field, as in the following example.

```
<field name="color">
   <grammar src="rgb.gram">
   <prompt>
      Please say red, green or blue.
   </prompt>
   <noinput>
      <prompt>
         I'm sorry, I didn't hear anything. Please
         say red, green or blue now.
      </prompt>
   </noinput>
   <nomatch>
      <prompt>
         I'm sorry, I didn't understand what you
         just said. Please say red, green or
         blue now.
      <prompt/>
   </nomatch>
<field>
```

When the browser triggers an event, it will search for an event handler for the current field. If the event handler is found, it will execute the handler, then wait again for user input.

The event handler will usually 1) explain the problem, and 2) encourage the user to speak the correct input. The *reprompt* tag instructs the browser to repeat the original prompt. The above example could be re-written like this.

```
<field name="color">
   <grammar src="rgb.gram">
   <prompt>
      Please say red, green or blue.
   </prompt>
   <noinput>
      <prompt>
         I'm sorry, I didn't hear anything.
      <prompt>
      <reprompt/>
   </noinput>
```

```
<nomatch>
   <prompt>
      I'm sorry, I didn't understand what you
      just said.
   <prompt>
   <reprompt/>
</nomatch>
<field>
```

If no handler is found, the browser will execute a default handler.
The default handler for *noinput* is defined by VoiceXML to be
equivalent to:

```
<noinput>
   <reprompt/>
</noinput>
```

The default handler for *nomatch* is a browser-dependent message
which might say something like "I'm sorry, I didn't understand
what you just said", followed by *<reprompt>*.

8.14   Tapered prompting

In the examples we have presented so far, there is a fixed prompt
for each field. Tapered prompting is a technique which allows the
prompt to vary each time an item is visited. Each form item and
event handler has a hidden counter (the "prompt counter") which
is set to one each time the form is visited, and incremented (one
added) each time the item is re-visited. The *count* attribute for a
prompt tag, specifies the minimum value that the counter must
have in order for the prompt to be played. If no *count* attribute is
specified, "1" is assumed. When there are multiple *prompt* tags, the
browser will select the tag with the highest *count* value that may
be played, i.e. the largest value of *count* which is less than or equal
to the item's current prompt counter.

The *count* attribute can also be applied to handlers such as
<noinput> and *<nomatch>*. When the field is initialized, the
counter is set to zero, one is added each time the appropriate
condition (*noinput* or *nomatch*) is reported.

Consider the following example.

```
<field name="color">
   <grammar src="rgb.gram">
   <prompt count="1">
      Please select a color by saying
      red, green or blue or blue now.
   </prompt>
   <prompt count="2">
      Say red, green or blue.
   </prompt>
   <noinput count="1">
      <prompt>
         I'm sorry, I didn't hear anything.
      </prompt>
      <reprompt/>
   </noinput>
   </noinput count="2">
      <prompt>
         I'm sorry, I didn't hear anything again.
         You may be speaking too quietly or
         waiting too long before speaking.
      </prompt>
      <reprompt/>
   </noinput>
   <nomatch count="1">
      <prompt>
         I'm sorry, I didn't understand what you
         just said.
      </prompt>
      <reprompt/>
   </nomatch>
   <nomatch count="2">
      <prompt>
         I'm sorry, I didn't understand that
         either. Please listen carefully to the
         following options and speak clearly.
      </prompt>
      <reprompt/>
   </nomatch>
<field>
```

A conversation based on this field definition might proceed as follows.

BROWSER *Entering field for the first time, so* *field prompt counter is 1:*	Please select a color by saying red, green or blue or blue now.
USER:	Orange.
BROWSER *Triggers nomatch event for the* *first time, so event handler* *prompt counter is 1:*	I'm sorry, I didn't understand what you just said.
BROWSER: *Executes <reprompt>, field* *prompt counter is now 2:*	Say red, green or blue.
USER:	Pink.
BROWSER: *Triggers nomatch event. Event* *handler prompt counter is now 2:*	I'm sorry, I didn't understand that either. Please listen carefully to the following options and speak clearly.
BROWSER: *Executes <reprompt>, field* *prompt counter is now 3:*	Say red, green or blue.
USER:	Blue.

As this example shows, the typical progression is towards shorter
prompts (the user has heard them before) and longer error
messages (the user apparently needs more help) as the count
increases.

9 VoiceXML Text-To-Speech Markup

9.1   Speech Synthesis Markup Language

The Speech Synthesis Markup Language (SSML) is an XML language for text-to-speech mark-up. It is fully embedded within VoiceXML. Historically, SSML derives from JSML, the Java Speech Markup Language.

Conceptually (and this is likely to be the real implementation), the VoiceXML browser hands off text to a synthesis engine which is a logically separate piece of software, possibly made by a different vendor. The text which is sent to be synthesized can be marked up using SSML tags. The browser hands off the marked-up text unmodified except for substituting any *<value>* tags.

We have already introduced some SSML tags including *<break>* for inserting a pause and *<voice>* for controlling some features of the synthesized speech, such as the age and gender.

The root tag for SSML is *<speak>*. This is the one tag in SSML which is not carried over directly into VoiceXML. In VoiceXML, text to be synthesized speech is specified by the *prompt* tag. (As we noted previously, the *prompt* tag may be omitted under some circumstances but doing so precludes additional markup and we don't recommend this practice).

All SSML tags affect only the text (and any child tags) they contain. Any settings they change are restored after the closing tag.

VoiceXML browsers are not obligated to fully implement all SSML tags, however, a compliant browser must silently ignore any tags it does not support, and should make its best efforts to map any unimplemented mark-up to something similar that is supported.

9.2 Sentence structure

The *sentence* and *paragraph* tags are used to explicitly delimit sentences and paragraphs. For example,

```
<prompt>
   <paragraph>
      <sentence>
         Good morning, ladies and gentlemen.
      </sentence>
      <sentence>
         Please welcome our guest speaker, Prof.
         Albert Einstein.
      </sentence>
   </paragraph>
</prompt>
```

The synthesizer uses these tags as hints on how to inflect the generated speech. The details will depend on the language For example, in English a sentence ending with a period (a simple declarative sentence) tends to end on a falling note, but a question ending in a question mark (an interrogatory sentence) typically ends on a rising note. The end of a paragraph indicates perhaps a longer pause and a different kind of emphasis at the beginning of the next sentence.

If explicit sentence and paragraph tags are not provided, the synthesizer should attempt to determine these boundaries from its knowledge of the conventions of the written language. For example, in European languages, the end of a sentence can be found from the appropriate punctuation marks. This is not always quite as easy as you might think. Here is one sentence with three periods.

```
<prompt>
   According to Prof. Einstein, pi is 3.1415.
</prompt>
```

The tags <s> and <p> are synonyms for <sentence> and <paragraph> respectively.

9.3   Language specification

The structural tags *sentence, paragraph, s* and *p*, the root tag of the
XML document, *vxml*, and *prompt* all support an optional attribute
named *xml:lang* which specifies the (human) language of the
document. For example,

```
<prompt xml:lang="en-US">
   Howdy, stranger.
</prompt>
<prompt xml:lang="en-GB">
   Jolly nice to meet you.
</prompt>
```

The format of the attribute value is a two-letter language code as
specified by ISO 639, such as "en" for English, followed optionally
by a hyphen and a sub-specifier, which may be an ISO 3166
country code, such as "US" for the United States or "GB" for Great
Britain.

Some example ISO 639 language codes are as follows.

AR	Arabic	NO	Norwegian
DA	Danish	RU	Russian
EN	English	SV	Swedish
ES	Spanish	TR	Turkish
FR	French	UK	Ukrainian
GB	United Kingdom	VI	Vietnamese
JP	Japanese	ZH	Chinese
KO	Korean	ZU	Zulu

And here are some ISO 3166 country codes.

AR	Argentina	FR	France
AT	Austria	GB	United Kingdom
AU	Australia	HK	Hong Kong
BR	Brazil	JP	Japan
CA	Canada	RU	Russian Fedn.
CN	China	SE	Sweden
DE	Germany	US	United States

It is easy to find complete listings by searching the Web for the ISO standard number.

Other sub-specifier types might give a dialect, for example *en-cockney* or *en-brooklyn*.

The browser determines a default language if none is specified. A child tag inherits the language from its parent, or can override it by specifying a different *xml:lang* value.

9.4   Interpretation

The *say-as* tag is used to specify how to interpret a text fragment which might be spoken in different ways. For example, "12" might be spoken as "twelve", "one two" or "twelfth". The following example speaks "12" in exactly those ways.

```
<prompt>
   <say-as type="number">
      12
   </say-as>
   <say-as type="number:digits">
      12
   </say-as>
   <say-as type="number:ordinal">
      12
   </say-as>
</prompt>
```

The *say-as* tag has a required *type* attribute which specifies exactly how to interpret the enclosed text. For example, *type="number"* says that the text is to be interpreted as a numerical value. There are optional format specifiers for some of these types which modify the way the value should be spoken. For example, *type="number:digits"* says that the value should be spoken as a string of digits "one two three" rather than a single value "one hundred twenty three". The supported types and format specifiers are listed in the following table.

Type	Format Specifier	Description
acronym		Speak by spelling out the letters, as in IBM or USA (not as a word, as in DEC or NASA).
address		Postal address.
currency		Money amount.
date	d	Day.
date	dmy	Day, month, year.
date	m	Month.
date	mdy	Month, day, year.
date	y	Year.
date	ymd	Year, month, day.
duration	h	Hours.
duration	hm	Hours, minutes.
duration	hms	Hours, minutes, seconds.
duration	m	Minutes.
duration	ms	Minutes, seconds.
duration	s	Seconds.
measure		Measurement.
name		Proper name of person, place ...
net		Internet URL.
number	digits	Speak as digit string.
number	ordinal	Speak as ordinal, i.e. "first", "second"...
telephone		Telephone number.
time	h	Hours.
time	hm	Hours, minutes.
time	hms	Hours, minutes, seconds.

When the format descriptor describes multiple fields, as in *date:my*, the fields are assumed to be separated by one non-numeric character. For example, the following might be spoken as "January, two thousand and three".

```
<prompt>
   <say-as type="date:my">
       1/3
   </say-as>
</prompt>
```

The following example shows some more types.

```
<prompt>
   Thank you for calling
   <say-as type="acronym">
      US
   </say-as>
   Friendly Bank. The balance in
   your account number
   <say-as type="number:digits">
      1234
   </say-as>
   on
   <say-as type="date:ymd">
      2002/12/25
   </say-as>
   was
   <say-as type="currency">
      $1.20
   </say-as>
</prompt>
```

This might be spoken as:

"Thank you for calling U.S. Friendly Bank. The balance in your account number one two three four on December twenty-fifth two thousand two was one dollar and twenty cents".

The *say-as* tag has an optional *sub* attribute. It provides alternative text to be spoken by the synthesizer; the text inside the tag is ignored. The following fragment would speak "I used to work for the World Wide Web Consortium".

```
<prompt>
   I used to work for
   <say-as sub="the World Wide Web Consortium">
      W3C
   </say-as>
</prompt>
```

The *sub* attribute is provided for text which may be marked up in more than one way — both for graphical display and for voice, for example. Since HTML browsers must ignore tags they don't recognize, this kind of scheme is sometimes useful.

9.5   Inserting a calculated value

The *<value>* tag is similar to the *<say-as>* tag, except that the value
to be interpreted is derived from an ECMAScript expression
instead of from the text enclosed by the tag. (ECMAScript is
described in a later chapter). The *expr* attribute gives the
expression. By default, the expression is evaluated as a string and
the string is inserted directly into the text. The following two
prompts are exactly equivalent.

```
<prompt>
   The quick <value expr="'brown'"/> fox.
</prompt>

<prompt>
   The quick brown fox.
</prompt>
```

The *class* attribute can be specified to control the interpretation of
the result. It can take exactly the same values as the *type* attribute
of *<say-as>*. For example,

```
<prompt>
   You are the <value expr="queue_pos"
     class="number:ordinal"/> caller
   in line, your call will be answered
   in approximately <value expr="est_time"
     class="duration:m"/>.
</prompt>
```

Some values can be spoken by concatenating pre-recorded sound
files. For example, a date such as "January seventh" can be played
using a vocabulary of 43 files (12 months and 31 ordinals). This
method is preferred when other prompts are also pre-recorded
and can be requested by setting the *mode* attribute to *"recorded"*
instead of the default value *"tts"*. A browser is free to ignore the
request and use speech synthesis anyway.

A browser is likely to support many different applications which
use different pre-recorded voices. To accommodate this situation,
the *audiobase* attribute is provided to tell the browser which

vocabulary to use (a vocabulary is a set of pre-recorded files containing words such as numbers and month names). The interpretation of this attribute is browser-dependent, but the typical use is expected to be to specify a base URL under which the browser can find the needed vocabulary files. The previous prompt might be re-designed to use pre-recorded sound files as follows.

```
<prompt>
   <audio src="/Audio/Fred/YouAreThe.wav"/>
   <value expr="queue_pos"
     class="number:ordinal" mode="recorded"
     audiobase="/Audio/Vocabs/Fred"/>
   <audio src="/Audio/Fred/EstTime.wav"/>
   <value expr="est_time"
     class="number:ordinal" mode="recorded"
     audiobase="/Audio/Vocabs/Fred"/>
</prompt>
```

9.6   Selecting a voice

You can request the age and gender of the generated voice by using the <voice> tag. The gender attribute is set to "male", "female" or "neutral". The age attribute can be used to give a numerical value in years, or you can set category to one of "child", "teenager", "adult" or "elder". All of these attributes are optional. For example,

```
<prompt>
   And now it's time for a story.<break/>

   <voice age="3" gender="neutral">
      Who's been eating
      <emphasis>my</emphasis> cereal?
   </voice>

   <voice category="adult" gender="female">
      Said Baby Bear.
   </voice>

   <voice category="elder" gender="male">
      Who's been eating
      <emphasis>my</emphasis> cereal?
   </voice>
```

```
<voice category="adult" gender="female">
   Said Grandfather Bear.
</voice>
</prompt>
```

The text inside a given *<voice>* tag is spoken with the requested attributes, the voice reverts to the previous settings when the end tag is reached.

9.7   Pronunciation

Fine control over pronunciation is provided by the *phoneme* tag. The pronunciation is specified by using a phonetic alphabet. The text inside the tag (if any) is ignored; however it is recommended that this be used to give a human-readable version of the words to be spoken. For example, the following specifies exactly how to say "tomato".

```
<prompt>
   You say
   <phoneme alphabet="ipa">
     ph="t&#x252;m&#x251;to&#x28A;">
     tomato
   </phoneme>.
</prompt>
```

The required *ph* attribute specifies the phonemes to be spoken. The optional *alphabet* attribute is set to *"ipa"* if the phonemes are given in the International Phonetic Alphabet (IPA), *"worldbet"*, to indicate the Worldbet alphabet, or *"xsampa"* for X-SAMPA.

In the above example, an IPA phoneme is specified as a Unicode character using the standard XML escape sequence "&#x*nnnn*;" where *nnnn* is up to four hexadecimal digits. Unicode characters in the range hex 250 to 2AF are reserved for IPA symbols, which include not only phonemes but a syllable delimiter, numerous diacritics, stress symbols, lexical tone symbols, intonational markers and more. Our personal favorite is 2AD, which is officially described as "audible teeth gnashing".

Experts consider all of these phonetic alphabets to be incomplete: they do not contain enough sounds to cover all human languages. Worldbet is considered the best in this regard, but IPA is probably the most widely used and understood.

9.8   Prosody

The technical term prosody corresponds roughly to the informal term inflection; it describes the shape or melody of speech, which is formed by changing speed, pitch, tone and so on. The *prosody* tag is used to set these parameters. For example,

```
<prompt>
   Listen carefully, the password is
   <prosody rate="slow" volume="soft">
      elocution.
   </prosody>
</prompt>
```

The attributes of *<prosody>* are shown in the following table. All are optional. If any attribute is omitted, this means "do not change", it does not mean "re-set to default".

Attribute	Values
pitch	The baseline pitch as a number (Hz), a relative change, or one of "high", "medium", "low", "default".
range	The pitch range (variation in pitch) as a number (Hz), a relative change, or one of "high", "medium", "low", "default".
rate	Speaking rate as an absolute number of words per minute, relative change, or one of "fast", "medium", "slow", "default".
volume	Absolute value in the range 0.0 to 100.0, a relative change, or one of "silent", "soft", "medium", "loud", "default".

Attribute	Values
duration	Length of time it should take to speak the text, a numerical value (optional decimals) followed by "s" for seconds or "ms" for milliseconds.
contour	See below.

Relative changes to values are specified as signed numbers with optional decimals, for example *volume="+12.5"*. The change can be expressed as a percentage, e.g. *volume="-10%"*.

Musically inclined readers might think of pitch in semitones and octaves. A semitone rise in pitch is approximately +5.9% and a semitone drop is -5.6%. A single tone (two-semitone) shift is +12.2% or -10.9%. A one-octave shift (12 semitones) is +100% or -50%, that is, the pitch doubles or halves. For pitch and range, the change can be expressed as semi-tones by using the "st" suffix, for example *pitch="+3st"*.

The typical baseline (average) pitch for a female voice is between 140 Hz and 280 Hz, with a pitch range of 80 Hz or more. Male voices are of course typically lower with a baseline of 70 to 140 Hz and a range of 40 to 80Hz.

A typical rate for reading aloud speech (such as a radio announcer) is 180 words per minute.

If both *duration* and *rate* are specified, *rate* is ignored. Similarly, *contour* takes precedence over both *pitch* and *range*.

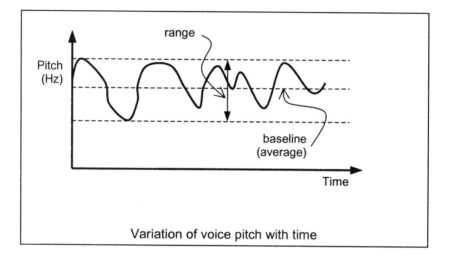

Variation of voice pitch with time

The *contour* attribute provides a way to describe how the pitch varies over time. It is specified by a series of (position, pitch) pairs. The position is specified as a percentage of total time needed to speak the text, given as a numerical value with optional decimal point followed by a percentage sign "%". The pitch can be specified using any value that would be valid for the *pitch* attribute. For example,

```
<prompt>
   <prosody contour="(0%,+40%)(50%,-80%)">
      Uh-oh!
   </prosody>
</prompt>
```

This raises the pitch 40% for the first half then drops the pitch 80% for the second half of the text "Uh-oh!". The synthesizer interpolates a smooth curve between the points given.

10 VoiceXML Grammars

10.1   Speech Recognition Grammar Format

The Speech Recognition Grammar Format (SRGF) is an XML language for writing voice recognition grammars. It is fully embedded within VoiceXML. Historically, SRGF derives from JSGF, the Java Speech Grammar Format.

As a historical note, the VoiceXML 1.0 specification made extensive use of JSGF in examples. However, JSGF was never required or even recommended for VoiceXML 1.0 browsers.

VoiceXML 2.0 compliant browsers must support SRGF. Other grammar formats may also be supported, depending on the browser.

There is an alternative form of SRGF which is not based on XML but on Augmented Backus-Naur Form (ABNF). Backus-Naur Form (and its Augmented and Extended brethren) are familiar to computer scientists as a way of representing programming language syntax.

The ABNF form of SRGF has the advantage that it may be easier for people to read and write. It also has the advantage that it may be possible (depending on browser support) to specify ABNF grammars in-line in a VoiceXML page, unlike the XML form which must always be in a separate file (at least, according to the current draft standard, this is likely to change). The capabilities of the two flavors are equivalent, and they are designed so that automatic translation between the two formats is straightforward.

Browsers may also support other grammar formats. The only format which is mandatory for a VoiceXML 2.0 compliant browser is the XML form of SRGF.

Proposed MIME types are "application/grammar+xml" for the XML form and "application/grammar" for the ABNF form.

Proposed standard filename extensions for SRGF files are ".grxml" for the XML form and ".grm" for the ABNF form.

10.2 Tokens

The fundamental building block of an SRGF grammar is a token. A token is the smallest unit of speech that can be directly matched by the recognizer. Typically a token is a single word.

Tokens are generally specified by spelling them in their native language. Here are some example tokens.

```
help
two
2
```

If a token consists of more than one word or contains non-alphanumeric characters, then it must be quoted using single or double quotes. These are also single tokens.

```
"James Roberts"
">"
```

The intent with the last token might be to match when the user speaks "greater than" — in which case it would probably be better to write out the words rather than using the symbol.

As an alternative to quoting, a token may be enclosed in a *token* tag:

```
<token>
   James Roberts
</token>
<token>
   greater than
</token>
```

The *token* tag has an optional attribute *xml:lang* (see the chapter on SSML for a discussion). It specifies the expected pronunciation. For example, in the US, "herb" is pronounced without the "h", it sounds like the "urb" in "suburb". In the UK, "herb" is pronounced as in the name "Herbert", with the "h". Assuming the grammar compiler is aware of this, the following two definitions should be different.

```
<token xml:lang="en-US">
   herb
</token>
<token xml:lang="en-GB">
   herb
</token>
```

A grammar can be as simple as a single token. Obviously, such a grammar is matched if the recognizer matches the token.

```
<grammar>
   help
</grammar>
```

The following two grammars are equivalent.

```
<grammar>
   "greater than"
</grammar>

<grammar>
   <token>
      greater than
   </token>
</grammar>
```

10.3 Sequences

A sequence can be built from two or more tokens. The following grammar has one sequence of two tokens.

```
<grammar>
   hello world
</grammar>
```

The tokens in a sequence must match in the order they are specified. The above grammar will match "Hello, world" but not "World, hello" or "Hello, cruel world".

A sequence may optionally be contained in an *item* tag. This is useful in more complex grammars where sequences may be embedded in other constructions. The following grammar contains a sequence of three tokens.

```
<grammar>
   <item>
      hello cruel world
   </item>
</grammar>
```

10.4 Choices

A choice is a set of alternatives, exactly one of which must match. The set of alternatives is contained within a *<one-of>* tag. Each alternative must be contained within an *<item>* tag.

The following grammar contains one choice with three alternatives.

```
<grammar>
   <one-of>
      <item>red</item>
      <item>green</item>
      <item>blue</item>
   </one-of>
</grammar>
```

This grammar will match "red", "green" or "blue".

10.5 Counts

The *<count>* tag provides a way to say that its contents are optional or may be repeated several times, as specified by its *number* attribute. The following table gives legal values for *number*.

Attribute	Meaning
number="?"	Optional, the contents may be present zero or one time in the utterance.
number="optional"	Same as *number="?"*.
number="0+"	The contents may be present zero, one or more times in the utterance.
number="1+"	The contents may be present one or more times in the utterance.

For example, the following grammar matches "Hello, world" and "Hello, cruel world".

```
<grammar>
    hello
    <count number="optional">
        cruel
    </count>
    world
</grammar>
```

Note that you can't write things like *number="3"* to indicate exactly three repeats, or *number="2-5"* to indicate two to five repeats. An exact number of repeats can be written as a sequence. A range, such as two to five, can be written as a *<choice>* where each item in the choice has a sequence of the appropriate length.

10.6   Rules

A rule is part of a grammar which is matched against speech. A token is the simplest possible rule. We have so far seen three ways to construct more complex rules: a sequence of tokens is a rule, a choice of tokens is a rule and a count of tokens is also a rule.

Sequences, choices and counts can be built from other rules as well as from tokens. You can have a sequence of choices, a choice of sequences, or more complex nested constructions.

The previous example showed a simple example of a nested construction where the grammar was built as a sequence *token, count, token.*

It is very convenient to be able to name a rule so that it can be used in multiple places. The *rule* tag gives a name to a rule definition without inserting the rule into the grammar. You might think of it as being like a function declaration: it gives some code which you can refer to elsewhere. The required *id* attribute gives the name. Here is a rule definition.

```
<rule id="primary_color">
   <one-of>
      <item>red</item>
      <item>green</item>
      <item>blue</item>
   </one-of>
</rule>
```

To insert a rule into the grammar, you use the *ruleref* tag. If *<rule>* is like a function declaration, then *<ruleref>* is like a function call. The name of the rule is specified by the *uri* attribute. A reference to a rule in the same page uses the *#name* notation (known as an anchor or fragment). A reference to the rule we just defined could look like this:

```
<ruleref uri="#primary_color"/>
```

You can also reference rules in other pages.

```
<ruleref uri="http://voice/grams.vxml#colors"/>
```

Note that SRGF does not permit more than one grammar to be specified per file, the fragment notation "#rulename" refers to a rule within a grammar, not a grammar within a file.

For example, here is a (slightly obscure) way to write a grammar to recognize "red", "green" or "blue".

```
<grammar>
   <ruleref uri="#primary_color"/>
   <rule id="primary_color">
      <one-of>
         <item>red</item>
         <item>green</item>
         <item>blue</item>
      </one-of>
   </rule>
</grammar>
```

By default, rule definitions are private, meaning that they can only be referenced by *ruleref* tags in the same grammar. The *scope* attribute of *<rule>* can be set to *"public"* to allow the rule to be referenced from other grammars.

10.7   Providing examples

The *rule* tag may contain one or more *example* tags which give examples of text which should match the grammar. The text in the example may be used as input for automated testing of the rule, documenting the rule for the human reader, or even for assisting the browser in generating useful error messages in the event of invalid input.

```
<grammar>
   <example>
      red
   <example>
   <one-of>
      <item>red</item>
      <item>green</item>
      <item>blue</item>
   </one-of>
</grammar>
```

10.8   Imports

An import is a convenience feature which assigns a name to a URL. The *import* tag is used to define the name, and it must precede any rule where the name is used. The *uri* attribute specifies the URL, the *name* attribute specifies the name. For example,

```
<import uri="http://voice/grams.gram"
  name="grams"/>
```

A rule in that page can now be referenced by using the *import* attribute of the *ruleref* tag.

```
<ruleref import="grams#colors"/>
```

The name may define a page, as in the above example, or a rule within a page.

```
<import uri="http://voice/grams.vxml#colors"
  name="colors"/>
```

Which can be referred to as:

```
<ruleref import="colors"/>
```

10.9   Special rules

SRGF defines three special rules: *#NULL*, *#GARBAGE* and *#VOID*. They are referenced like this:

```
<ruleref special="#NULL"/>
<ruleref special="#GARBAGE"/>
<ruleref special="#VOID"/>
```

The *#VOID* rule never matches. If a rule contains *#VOID*, it can never match.

The *#NULL* rule always matches. It does not "eat up" any input and does not change the set of utterances which matches a rule.

The *#NULL* and *#VOID* rules are mostly useful for testing and debugging browsers, recognizers and VoiceXML pages.

The *#GARBAGE* rule on the other hand is very useful: it matches anything or nothing. It is like the "*" or ".*" wildcard specifier in some regular expression and filename matching schemes. If there are one or more *#GARBAGE*s in a rule, then the recognizer will

match if there is any way of assigning partial utterances to
"garbage" that results in a match.

For example,

```
<grammar>
   I
   <ruleref special="#GARBAGE"/>
   fly
   <ruleref special="#GARBAGE"/>
   home
</grammar>
```

This will match "I want to fly home". The first *#GARBAGE*
matches "want to", the second matches nothing.

10.10 Rules in a grammar

If a grammar is composed of multiple rule definitions at the top
level, the grammar behaves as if they were a set of choices, in
other words the grammar will match any one of the rules. It's
important to be aware of this because you might expect it to
define a sequence, not a choice (but notice that we're talking about
a set of rule *definitions*, not a set of rules, so in the absence of this
convention such a grammar would in fact have no referenced
rules and would therefore be empty).

This can be overridden by specifying the *root* attribute of the
grammar tag. This says that the grammar should be treated as if it
contained one rule reference to the given rule. For example,

```
<grammar root="colors">
   <rule id="sizes" scope="public">
   ... etc...
   <rule id="colors" scope="public">
   ... etc...
</grammar>
```

A root rule must have public scope.

10.11 The ABNF form of SRGF

Support for the ABNF grammar format is optional in VoiceXML. It is certainly convenient for giving simple examples because ABNF grammars can be given in-line, unlike the XML form which must be specified in a separate file. We now give a quick overview of the syntax.

10.11.1 Tokens

Tokens are delimited by white space or by symbols with special syntactic function, such as the following.

```
; = | * + <> () [] {} /* */ //
```

Tokens may be explicitly quoted using single or double quotes if they contain white space or special symbols.

10.11.2 Comments

Comments may be specified using the C- and Java-like syntax. A comment either starts with "/*" and ends with "*/", or begins with "//" and continues up to the end of the current line.

10.11.3 Sequences

A sequence is written as a series of tokens and/or rule references. If necessary, parentheses (...) can be used to delimit the sequence. (Note that parentheses are never necessary in the XML form, the open and close tags play the role of open and close parenthesis). For example,

```
hello world
```

10.11.4 Choice

A choice is indicated by a vertical bar "|". A weight may be specified by giving a numerical value within forward slashes immediately before the item. Again, parentheses can be used for grouping. For example,

```
red | green | blue
( /0.3/ red | /0.2/ green | /0.5/ blue )
$digit | "#" | "*"
```

10.11.5 Counts

An optional item (zero or one times) is indicated by enclosing within square brackets [...]. Zero or more time is indicated by appending a star "*", one or more times is indicated by appending a plus sign "+".

```
[ please ]    // optional 'please'
$digit+       // one or more digits
```

10.11.6 Rules

A rule is named by putting one of the following three forms in front of a rule definition.

```
$rname =
public $rname =
private $rname =
```

Here, *rname* is the rule name. The rule definition is terminated by a semi-colon ";".

For example,

```
$color = ( red | green | blue);
$digit = (0|1|2|3|4|5|6|7|8|9);
```

10.11.7 Examples

Example text is specified by using a Javadoc-like commenting convention before the rule. For example,

```
/**
 * A primary color.
 *
 * @example red
 * @example green
 */
public $color = ( red | green | blue);
```

10.11.8 Rule reference

The following table compares the ABNF and XML form for rule references.

Reference type	ABNF syntax	XML syntax
In same page.	`$rname`	`<ruleref` `    uri="#rname/>`
In different page.	`URL$rname`	`<ruleref` `    uri=URL#rname/>`
In aliased page.	`$$aname#rname`	`<ruleref` `    import=aname#rname/>`
Root rule in aliased page.	`$$aname`	`<ruleref` `    import=aname/>`

Special rules.	`$NULL`
	`$VOID`
	`$GARBAGE`
		`<ruleref`
		`    special="#NULL"/>`
		`<ruleref`
		`    special="#VOID"/>`
		`<ruleref`
		`    special="#GARBAGE"/>`

Here, *rname* is a rule name and *aname* is an alias name.

10.11.9 Header declarations

The first non-comment declarations in a grammar file or in-line grammar should be the language, mode and root rule (in that order). All are optional. The syntax is as follows.

```
// An example ABNF grammar file
language en-US;
mode = speech; // speech or dtmf
root = color;
public $color = ( red | green | blue);
```

10.12 Semantic markup

In a VoiceXML field, when a user utterance is matched to a grammar, the field variable should be set to a value based on what the user said. For the built-in types, such as *digits*, it is well defined what the field variable should be. For user-specified grammars, however, the current draft of VoiceXML does not specify how the browser should set the value. In most of the examples in this book, we assume that the value of the variable is set to the text form of the user utterance and will therefore have the same value as the *utterance* property of the field's shadow variable (to be explained later). However, the VoiceXML specification does not specify that this is what should be done, at least in the draft current at the time of writing.

It would be very convenient to have more control over how values are extracted from grammars. There may be several utterances which the application would like to treat as the same thing. For example, the application might want to treat "Amex" in the same way as "American Express". A proposed solution for this, which may or may not be adopted in the final specification, is to add a *tag* attribute. (Note the confusing name—here "tag" is the name of an attribute). Given the following grammar, the field variable would be set to "Amex" whether the user said "Amex" or "American Express".

```
<rule id="CardType" scope="public">
   <one-of>
      <item tag="Amex">amex</item>
      <item tag="Amex">american express</item>
      <item tag="Visa">visa</item>
      <item tag="MC">mastercard</item>
      <item tag="MC">mastercharge</item>
   </one-of>
</rule>
```

The ABNF form attaches a tag within curly braces at the end of an item. Here is the ABNF form of the preceding example.

```
public $CardType =
   ( amex { "Amex" } |
   (american express) { "Amex" } |
   visa { "Visa" } |
   mastercard { "MC" } |
   mastercharge ( "MC" } );
```

A second use for semantic tagging is to extract the useful information out of a longer phrase, such as "My card is an Amex" or "I have a Mastercard". To accept any phrase ending with a recognized card type, we could modify our example grammar as follows.

```
<rule id="CardType" scope="public">
   <ruleref special="#GARBAGE"/> // match anything
   <one-of>
      <item tag="Amex">amex</item>
      <item tag="Amex">american express</item>
      <item tag="Visa">visa</item>
      <item tag="MC">mastercard</item>
      <item tag="MC">mastercharge</item>
   </one-of>
</rule>
```

Or, in the ABNF notation,

```
public $CardType =
   $GARBAGE
   ( amex { "Amex" } |
   (american express) { "Amex" } |
   visa { "Visa" } |
   mastercard { "MC" } |
   mastercharge ( "MC" } );
```

Without the semantic tags, the application would have to parse the utterance using ECMAScript in order to find a recognized sub-phrase.

Mixed-initiative forms, where a single grammar at the form level may fill in two or more fields, adds a second dimension to the problem. Now there are two questions: what value to assign to the variable, and what field does it belong to? At the present time, this issue is under study and solutions may be browser-dependent or may be defined in future revisions of the VoiceXML specification.

One proposal under consideration is to use ECMAScript expressions as semantic tags. This can be used to assign values directly to field variable names. Using this technique, the above example might look as follows.

```
<rule id="CardType" scope="public">
   <ruleref special="#GARBAGE"/> // match anything
   <one-of>
      <item expr="CardType='Amex'">amex</item>
      <item expr="CardType='Amex'">
        american express</item>
      <item expr="CardType='Visa'">visa</item>
      <item expr="CardType='MC'">mastercard</item>
      <item expr="CardType='MC'">
        mastercharge</item>
   </one-of>
</rule>
```

With this scheme, it is easy to deal with grammars which set more than one field. If you need to set more than one variable in one expression, you can use the ECMAScript comma operator, as follows.

```
expr="CardType='Amex',NeedDigits=16"
```

11 VoiceXML Scripting

11.1   ECMAScript in VoiceXML

A compliant VoiceXML browser must fully implement the scripting language variously known as JScript, JavaScript or ECMAScript. Officially, the language is defined by the ECMA-262 standard produced by the ECMA standards organization (www.ecma.ch) and adopted in June 1997. It is essentially identical to the ISO/IEC 16262 standard (www.iso.ch) which was based on ECMA-262 and approved in April 1998.

The language was invented by Brendan Eich of Netscape and was first supported by Netscape's Navigator 2.0 browser and later by Microsoft in Internet Explorer 3.0. All subsequent browsers from both companies have supported JavaScript (Netscape's name) or JScript (Microsoft's name), though with some significant version and dialect differences.

ECMAScript (as we shall call it here) is a powerful and quite complex language. It is beyond the scope of this book to give a complete tutorial or reference, however there are many books, Web sites and other resources available for the reader who needs a more elementary introduction or wishes to gain a more complete understanding. Our goal here will be to provide a solid understanding of the language sufficient for a typical VoiceXML developer. We assume that the reader is familiar with programming language concepts and preferably has some knowledge of C, C++ or Java.

Adding ECMAScript to VoiceXML allows the application to make sophisticated decisions and manipulate data on the client, for example to validate user input, without having to involve the server. There are no functions which allow access to external APIs such as the browser file system, databases etc. (with one very minor exception: you can get the current date and time from the browser's system clock — remember, the browser's clock may be out of sync with the server!).

ECMAScript appears in VoiceXML inside a *script* tag and in the *cond* and *expr* attributes.

The *expr* attribute is supported by several tags. It allows you to give a script expression in place of an attribute such as *next* or *src*. To give a trivial example, the following two *<audio>* tags are exactly equivalent (apart perhaps from a slight performance penalty for loading the script interpreter).

```
<audio src="HelloWorld.wav"/>
<audio expr="'HelloWorld.wav'"/>
```

In the *audio* tag, the *expr* attribute may be used to specify the URL of the sound file as an alternative to the *src* attribute. As you would expect, *'HelloWorld.wav'* is a valid ECMAScript expression which yields the string *HelloWorld.wav* as a result.

The *cond* attribute provides an expression which is evaluated as a boolean true / false result which determines whether or not to execute the action of the tag. For example,

```
<prompt>
    You win.
</prompt>
<prompt cond="wins > 6">
    Wow, you're really good at this.
</prompt>
```

The *script* tag contains a block of script code. It is similar to *<script>* in HTML, except that there is no *type* attribute: the language must be strictly identical to the language defined by ECMA-262, no variation allowed. This is of course intended to improve the portability of VoiceXML compared with HTML.

There are two main uses of *<script>*: defining functions which can be called from *expr* and *cond* expressions, and running scripts to compute variable values to be used elsewhere in the page. The *script* tag can appear in executable content and as a child of the *<vxml>* tag.

To create an example for *<script>* we'll introduce the *value* tag, which allows you to insert variables such as numbers into a prompt. For example,

```
<prompt>
    You have <value expr="5"/> new messages.
</prompt>
```

This will speak "You have five new messages". We will discuss *<value>* in more detail later.

Our example in the following page defines a trivial function *square* which computes the square of its argument and uses this function in a *<value>* tag. The user is prompted for a number; the browser speaks back its square. (The example could be simplified by simply computing the square directly inside the *expr* attribute, but this allows us to show a function without introducing too many new concepts).

```
<?xml version="1.0"?>
<vxml version="2.0">
   <script>
      function square(x)
         {
         return x*x;
         }
   </script>
   <form>
      <field id="n" type="number">
         <prompt>
            Please say a number.
         <prompt>
      </field>
   </form>
   <block>
      <prompt>
         The square of <value expr="n"/>
         is <value expr="square(n)"/>.
      </prompt>
      <disconnect/>
   </block>
</vxml>
```

This page defines a form with two items. The first is a field named *"n"* which collects a number, the second is a *<block>* which confirms the value, speaks its square, and hangs up. This example uses the fact that each named field has a corresponding ECMAScript variable with the same name as the field which contains the value collected from the user. So, in this case, there is an ECMAScript variable *n* which contains the number spoken or dialed by the user.

The *script* tag may be used wherever executable content is allowed or as a child of the root tag, *<vxml>*. In this example the script is run when the page is loaded and before the form is interpreted. Running this particular script defines a function but does not call it, the function code is not executed until it is called from an expression elsewhere.

11.2   Data types and constants

There are three fundamental types in ECMAScript: Numbers, Strings and Booleans.

11.2.1 Numbers

Numbers are always floating point; there are no integer types. The internal format used is (or must produce results equivalent to) double-precision 64-bit IEEE 754 values including the special values "Not-a-Number" (*NaN*), positive infinity (*Infinity*), and negative infinity (*-Infinity*). Number constants are represented using decimal digits, an optional decimal point and an optional decimal exponent preceded by "e".

Integer values (meaning of course floating point values with zeros to the right of the decimal point) may be specified using the same notation as Java or C: as decimal digits, as octal digits (if the first digit is zero), or as hexadecimal digits (by using the 0x prefix).

Following are some legal numerical constants.

```
123        Decimal
0123       Octal
0x123      Hex
123.45
1.6e10
-7
.2
-.2e-17
Infinity
-Infinity
NaN
```

11.2.2 Strings

An ECMAScript string is an ordered sequence of 16-bit Unicode characters. String constants may be written in single or double quotes. A double quote may appear in a single-quoted string and vice-versa. Inside a string, the backslash character "\" is reserved for the escape sequences shown in the following table.

Escape	Character
\b	Backspace
\t	Tab
\r	Carriage-return
\n	Newline
\t	Tab
\ '	Single quote
\ "	Double quote
\\	Backslash
\uxxxx	Unicode character with hex value *xxxx*.

Following are some legal strings.

```
"3"
"Didn't you hear about VoiceXML?"
'No, I didn\'t'
"\"\\\u0020\" is backslash space"
```

11.2.3 Booleans

An ECMAScript boolean value is a logical value represented by the constant values *true* and *false*.

11.3 Comments

Comments may be included in scripts using the same syntax as C. Any characters between /* ... */, or between // and the end of the line, are ignored.

11.4 Identifier names

Identifiers, in other words names used for variables, functions etc., are one or more characters from the following set:

letters (A-Za-z), digits (0-9), underscore (_), dollar ($).

The first character may not be a digit. The dollar sign is intended for use only in machine-generated code. Following are legal names.

```
AName
x2345
$Temp
_x
a_longer_name
```

As with everything else in ECMAScript, names are case sensitive, so *name* and *Name* are different.

11.5 Variables

A variable is declared by using the *var* statement. For example,

```
var x;
```

An initial value may optionally be given.

```
var x = 1;
```

The initial value may be specified by any valid expression, which may include other variables.

```
var y = 1;
var x = y + 2;
```

If no initial value is given, the variable is set to a special value called *undefined*. So the first example above is exactly equivalent to:

```
var x = undefined;
```

It is legal to "re-declare" a variable that already exists, in which case the *var* statement is treated as an assignment.

There is another special value called *null* which indicates that the variable holds no value (as opposed to being uninitialized).

Variables should be declared before they are used in a VoiceXML script or expression.

Variables do not have a fixed data type, however the variable stores both a value and the data type of that value. Visual Basic programmers will recognize this as being similar to *Variant* variables, COM programmers will see an analogy with the *VARIANT* structure.

The *typeof* operator may be used to find the type of the data currently stored in a variable. The returned value is a string representing the type name, which is *"number"*, *"string"* or *"boolean"* for the fundamental types. Consider the following example, in which the variable *x* changes its data type.

```
var x = 1;
var Type1 = typeof(x);   // Type1="number"
x = "2";
var Type2 = typeof(x);   // Type2="string"
x = true;
var Type3 = typeof(x);   // Type3="boolean"
```

Variables declared within a function may be used only within that function. Variables declared outside of a function may be used anywhere within the script.

11.6 Layout

As the preceding examples illustrate, ECMAScript statements are terminated by a semi-colon. It is sometimes legal to omit the semi-colon, but it is much safer to get into the habit of always using one.

White space is not significant except inside string constants or where needed as a separator (*varx* is different from *var x*).

11.7 Expressions

Expressions may be built from constants, variables, parentheses for grouping and operators which will mostly be familiar to C and Java programmers.

Unfamiliar operators, in addition to *typeof* which we have already described, include === and !==. These operators compare for equality and inequality respectively without performing type conversions, so values of different types are always considered not equal. The operators == and != perform type conversions to make their operands compatible before comparing.

This is illustrated by the following expressions, which both evaluate to *true*.

```
123 == "123"   // LHS is converted to string first
123 !== "123"  // Different, data types not same
```

11.8 Operators

The following table is a complete list of operators in precedence order. The precedence (*Pr* column) indicates the relative strength of an operator when found in an expression with other operators. For example, in the expression 2+3*4, multiplication is stronger than addition ("has higher precedence"), so will be performed first even though it appears later in the expression. In other words, this expression will be evaluated as *2+(3*4)*. The precedence is given as a number from 15 (strongest) to 1 (weakest).

The *A* column indicates whether the operator is left- or right-associative. The associativity of an operator specifies whether a sequence or two or more of the same operator is evaluated from left-to-right (L) or from right-to-left (R). Arithmetic operations such as addition are left-associative, so *2+3+4* is evaluated as *(2+3)+4*. Assignment operators are right-associative, so *x=y=z* is evaluated as *x=(y=z)*.

Unless changed by precedence or associativity, an expression is evaluated from left to right.

The *n* column indicates how many operands the operator takes: 1 for unary (such as logical not, "!"), 2 for binary (such as addition, "+") and 3 for tertiary (the conditional operator "?:").

Groups of operators with the same precedence are grouped into shaded bands.

Operator	Pr	A	n	Description
.	15	L	2	Select a property from an object (same as selecting an array element).
[]	15	L	2	Select a property from an object (same as selecting an array element).
()	15	L	1	Invoke a function (not the same as parentheses used for grouping).
++	14	R	1	Increment the operand (as in *x = x+1*). If written ++*x* then add on before using the value in rest of the expression, if *x*++ use the value then add one.
--	14	R	1	Decrement the operand (as in *x = x-1*). If written --*x* then add one before using the value in rest of the expression, if *x*-- use the value then add one.

Operator	Pr	A	n	Description
-	14	R	1	Change sign.
~	14	R	1	Bit-wise not (convert operand to 32 bit integer first).
!	14	R	1	Logical not.
delete	14	R	1	Delete a property.
new	14	R	1	Create a new object.
typeof	14	R	1	Result is operand's type as a string.
void	14	R	1	Coerce expression to *undefined*.
*	13	L	2	Floating-point multiplication.
/	13	L	2	Floating-point division.
%	13	L	2	Integer modulo.
+	12	L	2	If one operand is a string, convert the other operand to a string and concatenate. Otherwise, floating-point addition.
-	12	L	2	Floating-point subtraction.
<<	11	L	2	Bitwise left shift (convert left operand to 32-bit integer first).
>>	11	L	2	Bitwise right shift (convert left operand to 32-bit integer first, extend with sign bit).
>>>	11	L	2	Bitwise right shift (convert left operand to 32-bit integer first, extend with zero bit).
<, <=, >, >=	10	L	2	Floating-point numerical comparisons.

Operator	Pr	A	n	Description
==	9	L	2	Test for equality, perform type conversion if needed.
!=	9	L	2	Test for inequality, perform type conversion if needed.
===	9	L	2	Test for inequality, different data types always false.
!==	9	L	2	Test for inequality, different data types always true.
&	8	L	2	Bit-wise AND (convert both operands to 32-bit integers first).
^	7	L	2	Bit-wise XOR (convert both operands to 32-bit integers first).
		6	L	2	Bit-wise OR (convert both operands to 32-bit integers first).
&&	5	L	2	Logical AND, evaluate right-hand-side only if left-hand-side is *true*.
||	4	L	2	Logical OR, evaluate right-hand-side only if left-hand-side is *false*.
? :	3	R	3	Written *1st ? 2nd : 3rd*. Result is 2nd operand if 1st operand is *true* (3rd operand is not evaluated), otherwise result is 3rd operand (2nd operand is not evaluated).

Operator	Pr	A	n	Description
=	2	R	2	Assignment. Left-hand-side must be a variable. Result is right-hand-side.
op=	2	R	2	Assignment. *x op= y* is equivalent to *x = x op y*, where *op* is any binary numerical or bit-wise operator.
,	1	L	2	Sequence. The left-hand-side is evaluated for any side effects and discarded, result is the right-hand-side.

Note that the + operator (and its cousin, +=) gives different results depending on the data types of its operands. If one or both operands is a string, it will perform string concatenation, otherwise it performs numerical addition.

11.9 Statements

Again, the range of statements supported by ECMAScript will be very familiar to C and Java programmers.

The simplest form of a statement is an expression followed by a semi-colon. The typical example is an assignment, such as:

```
x = 1;
```

Compound statements may be formed by including two or more statements in curly braces { ... } . This allows more than one statement to be controlled by an *if*, for example.

Conditional statements are built using *if* , which can take either of the following forms:

```
if (condition)
    statement

if (condition)
    statement
else
    statement
```

A *statement* is an expression which is evaluated as a boolean true / false value. Here is an example of an *if*-statement.

```
if (x > 2)
    {
    y = 3;
    n = 0;
    }
else
    y = 4;
```

Loops of the usual *for*, *do...while* and *while* types are supported. The syntax for loop statements is as follows.

```
for (initialize ; test ; increment)
    statement

do
        statement
while (test);

while (test)
    statement
```

Here, *test* is a condition which is evaluated every time around the loop, if it evaluates to *false* the loop is terminated; *initialize* is an expression which is evaluated once before the loop is entered, and *increment* is an expression which is evaluated each time the statement in the loop body has been fully executed.

The *break* and *continue* statements work in the same way as in C and Java: they break out of the loop and jump to the start of the next iteration of the loop respectively. Labeled versions are supported as in Java.

A special form of the *for* loop iterates over the properties of an object:

```
for (variable in object)
    statement
```

Suppose that *obj* is a reference to an object which currently has properties *a, b* and *c*. Consider the following example.

```
var v;
for (v in obj)
    f(v);
```

This is equivalent to:

```
var v;
v = obj.a;
f(v);
v = obj.b;
f(v);
v = obj.c;
v(v);
```

We will explain more about objects and properties shortly.

The *switch* statement again works very much like C and Java. The syntax is:

```
switch (swexpr)
    {
case caseexpr:
    statements
case caseexpr:
    statements
    . . .
default:
    statements
    }
```

The expression *swexpr* is evaluated and then compared one by one with the case expressions *caseexpr* until one matches. Comparison is done as if the === operator was used, so type conversions are not performed. When a match is found, execution branches to the first statement following the *case* label. As in C and Java, execution will fall through to the next case unless *break* is used. An optional *default* label may be specified, execution will branch there if there is no match to any case.

As in Java, there is no *goto* statement.

ECMAScript does support exceptions through *try, catch, finally* and *throw.* Exceptions can be used for both language errors (e.g., trying to interpret an illegal statement such as *1=2;*) and for user-defined errors. Further discussion of exceptions is beyond the scope of this book.

11.10 Functions

Here is a simple function.

```
function Add(x, y)
    {
    return x + y;
    }
```

It starts with the keyword *function,* followed by parentheses containing zero or more argument names. The body of the function is enclosed in curly braces *{ ... }.* The *return* statement may be used anywhere within the function to return a value. If there is no *return* statement, the function returns no value (in some languages this would be called a subroutine).

Function parameters are passed by value.

Unlike C++ and Java, ECMAScript does not support function name overloading: there may only be one function with a given name. If more than one is found in the source code, the most recent definition is used.

Functions may be called with a different number of arguments than the number in the function definition. An argument which is in the definition but not supplied in the call gets the value *undefined.* Extra arguments must be obtained using the *arguments* object (explained later).

A function definition is also considered a statement and may appear anywhere a statement would be valid. At run-time, execution simply skips around a function definition, it behaves like an empty statement (a semi-colon). Functions may be defined inside functions. Forward references, i.e. calls to functions which are defined later in the same script, are permitted.

Unlike C and Java, there is no concept of a *main* function; execution simply starts at the first statement in the script which is not inside a function.

11.11 Objects

While everything so far has been mostly old hat or an easy transition for C and Java programmers, ECMAScript objects and classes will be quite unfamiliar.

An object is created by using the *new* operator. The right-hand-side of *new* is a constructor expression. We'll expand on constructors in a moment, but there is a built-in constructor for a minimal object called *Object()* which we can use to get started.

```
var obj = new Object();
```

As this illustrates, in addition to the fundamental types, a variable can contain a reference to an object. A reference is similar to a pointer, when it is copied it is only the address or some other internal handle to the object which is copied, not the data in the object itself.

The new object has no properties, but you can add some simply by assigning them, as follows.

```
obj.Property = 1;
obj.AnotherProperty = "A";
```

An alternative to the *objectname.propertyname* notation is the array index syntax *objectname["propertyname"]*, so this is equivalent to the previous example:

```
obj["Property"] = 1;
obj["AnotherProperty"] = "A";
```

ECMAScript objects can be considered associative arrays where the string name of a property is mapped to a variable value.

Objects can have properties added to them at any time. A given property of an object behaves in most respects like a variable. A property can be removed from an object by using the *delete* operator.

```
delete object.AnotherProperty;
```

Objects cannot be explicitly deleted; it is up to the interpreter to free objects when they are no longer used.

11.12 Classes

To create a class, you start by defining a function which acts as the constructor for that class. Within that function you can use the special variable called *this* to refer to the object being created. As a simple example, we'll create a class called *Point* which has two properties x and y (imagine that they store the x,y coordinates of a point in a plane). The constructor function could look like this.

```
function Point()
   {
   this.x = 0;
   this.y = 0;
   }
```

If this function has been defined, we can create a new object as follows.

```
var pt = new Point();
```

Now *pt* contains a reference to an object with two properties named x and y, both of which have been set to zero.

We can define an argument list for the constructor function if we wish.

```
function Point(initX, initY)
  {
  this.x = initX;
  this.y = initY;
  }
```

This is used as follows.

```
var pt = new Point(1, 2);
```

Note that you can't define more than one constructor for a class (this is consistent with the rule that you can't have more than one function with the same name).

You can add methods to a class by assigning function names to properties. Suppose we want to add a *Dist* function to our *Point* class which returns the distance of the point from the origin. First we define *Dist* as follows.

```
function Dist()
  {
  return Math.sqrt(this.x*this.x +
   this.y*this.y);
  }
```

The *Math.sqrt* function is built-in to ECMAScript; as you would guess it returns the square root of its argument. Given this function, we can expand our constructor as follows.

```
function Point(initX, initY)
  {
  this.x = initX;
  this.y = initY;
  this.Dist = Dist;
  }
```

Now we can use the *Dist* method as follows.

```
var pt = new Point(3, 4);
var d = pt.Dist();    // d=5
```

11.13 Classes

ECMAScript defines a number of built-in classes, as described in the following table.

Class	Description
Arguments	Inside a function, a special object named *arguments* is available which contains the arguments actually passed to the function (there may be more or fewer than the function definition). For example, *arguments.length* contains the number of actual arguments, *arguments[n]* contains the n'th argument where n = 0, 1, ... length-1.
Array	An array of elements addressed using integer indexes. Provides functions for combining, splitting, sorting and expanding arrays.
Boolean	The constructor provides a convenient way to convert non-Boolean values to *true* or *false* in places where a boolean value is mandatory.
Date	Provides a wide range of functions for manipulating dates.
Error	Error objects are thrown by exceptions.
Function	Treats a function as an object. Less efficient but more flexible than defining a function in code.
Global	A placeholder for certain global properties. You can't create an object of this type.
Math	A placeholder for a range of useful mathematical functions. You can't create an object of this type.
Number	A placeholder for a number of useful read-only constants. For example, Number.MAX_VALUE is the largest possible floating-point value.
Object	A minimal object.

Class	Description
RegExp	Provides functions for using regular expressions (a powerful pattern-matching and search-and-replace technique for strings).
String	Provides a set of methods for manipulating strings, e.g. *substring*.

It is beyond the scope of this book to give a comprehensive description of these built-in classes.

11.14 The eval function

ECMAScript has a built-in function called *eval* which executes the code passed to it through a string argument. If the code is an expression, the return value from *eval* is the value of the expression. If the code is one or more statements, the statements are executed and the return value is *undefined*. For example,

```
var x = eval("2 + 2");
eval("if (n > 3) y = 1; else y = 2;");
```

This allows you to perform some interesting, though often obscure, programming tricks. Using *eval* is not recommended unless there is no reasonable alternative.

11.15 Experimenting with ECMAScript

Most graphical browsers support ECMAScript or something very close to it. Using a graphical browser makes it easy to edit and test. Type HTML with embedded script into any text editor and use a *file://...* URL to load the page into the browser. When you make a change, simply hit the *Refresh* button to re-run the script. The *document.write* function (supported in graphical browsers but not VoiceXML browsers) gives you an easy way to display results; you may find this to be a more convenient learning environment than a voice browser.

The following HTML page provides a skeleton within which you can try out ECMAScript programs.

```
<HTML>
   <SCRIPT language="JavaScript">
      document.write("Hello from ECMAScript");
   </SCRIPT>
</HTML>
```

This page should display the text "Hello from ECMAScript".

Here is a page which runs one of our earlier ECMAScript examples.

```
<HTML>
   <SCRIPT language="JavaScript">
      var x = 1;
      var Type1 = typeof(x);   // Type1="number"
      x = "2";
      var Type2 = typeof(x);   // Type2="string"
      x = true;
      var Type3 = typeof(x);   // Type3="boolean"
      document.write("Type1=" + Type1 +
        " Type2=" + Type2 + " Type3=" + Type3);
   </SCRIPT>
</HTML>
```

11.16 Embedding ECMAScript in a VoiceXML page

XML reserves the "<" and "&" characters. In an attribute value, either the single quote or double quote is also reserved (depending of course on whether you chose to enclose the value in single or double quotes). You must therefore be careful when using symbols such as "<" for less-than in an ECMAScript expression. When the script expression appears in an attribute value, you have no choice but to use an escape sequence such as "<". Inside *<script>*, you have the alternative of using a CDATA section. Using CDATA has the important restriction that you must avoid the three-character sequence "]]>" in the script; there is no way to represent it with an escape. If you can be certain that "]]>" does not appear, then using CDATA has the advantage that the script source code may be more readable because escapes like "<" are not needed.

11.17 Using script variables in VoiceXML

The *var* tag is used to declare a variable:

```
<var name="pi" expr="3.1415926535"/>
```

This is similar to the ECMAScript statement:

```
var pi=3.1415926535;
```

The required *name* attribute gives the name of the variable, the optional *expr* attribute gives the initial value. If no initial value is given, the variable is set to *undefined*. As in ECMAScript, it is legal to declare a variable which already exists, in that case the *var* tag is treated as an assignment.

Any legal ECMAScript name may be used, however names beginning with an underscore "_" or ending in a dollar sign "$" are reserved for internal use by the browser.

The variable will be re-initialized each time the parent tag is interpreted. For example, if the *var* tag appears as a child of <vxml>, it will be initialized once only when the page is loaded. If it is a child of a *form* tag, it will be initialized each time the form is processed.

Variable initialization takes place in the order the *var* tags are written in the source code (in so-called document order).

The *assign* tag assigns a value. It is very similar to the *var* tag, except that the *expr* attribute is required.

```
<var name="n"/>
<assign name="n" expr="2"/>
```

It is illegal to use *assign* for a variable that does not exist. Here VoiceXML differs from ECMAScript, where an assignment statement for a variable which does not exist is treated as a declaration (though we recommend that this should be avoided as bad style).

11.18 Variable scopes

The scope of a variable is the range in which the variable name is known and may be referenced.

Variables declared by a *var* statement inside a script (i.e., inside a *script* tag) are valid only within that one tag. Their values do not persist, they cannot be referenced from other scripts or expressions.

Variables declared by a *var* tag have a scope which extends anywhere within the parent tag of that <var>.

The following table summarizes the various scopes a variable may have.

Scope name	Parent of <var>	Description
session	*N/A*	Pre-defined variables such as *session.telephone.ani*.
application	<vxml>	Defined directly under *<vxml>* in the application root document. Initialized when root document is loaded.
document	<vxml>	Defined directly under *<vxml>* in the current page. Initialized when page is loaded.
dialog	<form>	Defined directly under *<form>*. (Menus cannot declare variables). Initialized when form is visited. Includes field variables.
item / *handler*	<block>, <filled>, <catch>	This is also called the "anonymous" scope. Tags which are shorthands for *<catch>*, such as *<noinput>*, also have this scope.

Some variables are defined by the browser. These are known as session variables. For example, *session.telephone.ani* contains the ANI digits (if known). Session variables are read-only. New session variables cannot be created by a VoiceXML page.

The application root document will be explained later. (It is a second VoiceXML page that remains loaded while the primary page is swapped out due to hyperlinking).

The scopes form a nested hierarchy which reflects the structure of a VoiceXML page, as the following diagram illustrates.

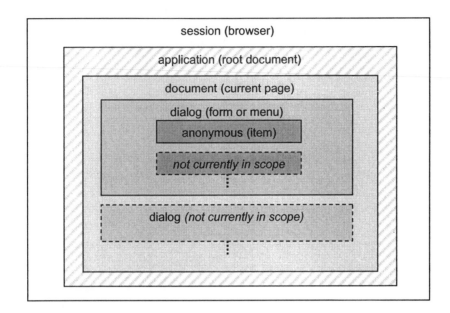

Expressions can refer to variables in an outer scope. If the outer variable is obscured by a variable of the same name in an inner scope, the outer variable can be referenced by using the name of that scope as a prefix. For example, an expression in dialog scope can refer to a session variable named *x* as *session.x*. This is because the browser provides "scope objects" named *session*, *application*, *document* and *dialog*. The scope object has one property for each variable within its scope. This property is a reference to that variable and has the same name, it is therefore a way to get and set the variable's value. A given scope object is visible in all scopes which are outside it in the hierarchy, so for example the document scope object is accessible from dialog scope but not from application scope.

The innermost scope is anonymous (meaning, it has no corresponding scope object) because there is no nested scope which might need to refer to one of its variables.

11.19 Session variables

The browser defines a standard set of variables at session scope. For example, *session.telephone.ani* contains the ANI digits captured at the beginning of the call, or *undefined* if there was none.

11.20 The <if>, <else> and <elseif> tags

Conditional code can be created in executable content by using *<if>*, *<else>* and *<elseif>*. The usage for *<if >*and *<else>* should be almost self-explanatory. The following example shows one way of avoiding an incorrect use of the plural, as in "one widgets".

```
<if cond="n==1">
   <prompt>
      You have one widget.
   </prompt>
<else/>
   <prompt>
      You have <value expr="n"/> widgets.
   </prompt>
</if>
```

The *<elseif>* tag also has a condition and provides for a multi-way decision. The browser first tries the *if*, then each *elseif* in the order written, until a *true* condition is encountered. The executable code between the matched condition and the next *<elseif>*, *<else>* or *</if>* ("endif") is run, execution then skips past the *</if>*. Here we extend the previous example to treat zero widgets as another special case.

```
<if cond="n==0">
   <prompt>
      You have no widgets.
   </prompt>
<elseif cond="n==1"/>
   <prompt>
      You have one widget.
   </prompt>
<else/>
   <prompt>
      You have <value expr="n"/> widgets.
   </prompt>
</if>
```

The *<elseif>* tag can be used to play a role like a *switch* statement in other languages, as in the following.

```
<if cond="blouse_color == 'red'">
   <prompt>
       We suggest a pink hat.
   </prompt>
<elseif cond="blouse_color == 'green'"/>
   <prompt>
       We recommend our pearl necklace.
   </prompt>
<elseif cond="blouse_color == 'blue'"/>
   <prompt>
       Our gold purse would match perfectly.
   </prompt>
<else/>
   <prompt>
       English white socks would be just the thing.
   </prompt>
</if>
```

11.21 Validating user input

One of the most important uses of scripting is to provide validation of user input that goes beyond matching to a grammar.

The *filled* tag contains executable content to be run when a field variable has been "filled in" by user input. If the input is found to be invalid, the typical action to take is to use the *clear* tag, which discards the user input and re-sets the field variable to *undefined*. This will have the result of causing the browser to make a new attempt to prompt and get input for that field. Suppose that we want to prompt the user for a number which is even, this can be verified as in the following example.

```
<field id="even_number" type="number">
   <prompt>
      Please say or dial an even number now.
   </prompt>
   <filled>
      <if cond="even_number%2 != 0"/>
         <prompt>
            Sorry, <value expr="even_number"/> is
            an odd number.
         </prompt>
         <clear/>
      </if>
   </filled>
</field>
```

If *clear* is the child of a form rather than a field, it clears all fields in the form, unless the *namelist* attribute is included to give a specific list, as in:

```
<clear namelist="even_number"/>
```

You can also use *filled* as a child of a form, in which case it can be triggered when a given set of the form's fields has been filled in. By default, all fields must be filled. You can also specify the *namelist* attribute, which gives a whitespace-separated list of field names which must be filled. For example, suppose a form has three fields named *card_number*, *card_type* and *exp_date*. Then *filled* with no attributes is equivalent to:

```
<filled namelist="card_number card_type exp_date">
```

The optional *mode* attribute is set to *"all"* to specify that all the fields must be filled in (the default), or *"any"* to specify that it should trigger if any one of the fields is filled in.

12 VoiceXML: Other Features

12.1 Shadow variables

For each dialog item such as a *field* tag, the browser creates an ECMAScript variable called the shadow variable. The browser sets properties of the shadow variable to provide information about the interpretation of the item which is not available through other means. The shadow variable name is constructed from the item name with a dollar character "$" appended.

For example, a field named *color* will create a shadow variable named *color$* with properties *color$.confidence, color$.utterance* and *color$.inputmode*.

The *confidence* property contains the confidence level of the match represented as a number between 0.0 (no confidence at all) to 1.0 (certainty). The *utterance* property contains the raw string of words matched by the recognizer. The exact format of the string will be browser-dependent. The *inputmode* property is set to "*dtmf*" to indicate that input was obtained by touch-tones, or "*voice*" or "*speech*" to indicate that input was obtained from voice recognition. The two values "*speech*" and "*voice*" are synonyms, but will be set to match the value specified by the *inputmode* attribute, for example with *<grammar mode="speech">*, "*speech*" is returned. Built-in grammars return "*voice*".

The following example forces a *nomatch* event to be thrown if the confidence level is less than 0.75, overriding the built-in default of 0.5. Note that the ECMAScript operator "<" for less-than must be escaped as "<". Note also that changing the confidence level would be better implemented by setting the *confidencelevel* property (properties are explained later).

```
<field name="color">
   <grammar>
      red | green | blue
   </grammar>
   <prompt>
      Please say red, green or blue now.
   </prompt>
```

```
        <filled>
           <if cond="color$.confidence &lt; 0.75">
              <throw event="nomatch"/>
           </if>
        </filled>
     </field>
```

12.2 Recording

The *record* tag is used to record audio. It is another special type of form item. The recording is stored in an ECMAScript variable, where it is available for playback (by using in a *value* tag) or for storing on the server (by using <*submit*>).

Here is a form which prompts the user to leave a message and then saves it on the server.

```
   <form>
      <record name="msg" beep="true"
         maxtime="120s" finalsilence="3s"
         dtmfterm="true" type="audio/wav">
           <prompt>
              Please record your message at the beep.
           </prompt>
      </record>
      <block>
           <submit next="SaveMsg.jsp" method="post"/>
      </block>
   </form>
```

Note the use of *method="post"* to submit the data, you should never use the default *GET* method to send an audio recording since this is almost certain to overflow the maximum length of URL permitted by your server.

The attributes of <*record*> are pretty self-explanatory. The *beep* attribute is set to *"false"* (the default) or *"true"* to indicate whether a beep should be played before starting to record, *maxtime* gives the maximum time to allow for recording, *finaltime* gives the amount of silence time that will terminate the recording, *dtmfterm* is set to *"true"* (the default) or *"false"* to indicate whether a touch-tone will terminate the recording. The *type* attribute optionally specifies the media type (sometimes called MIME type) of the audio data. If *type* is not specified, a browser-specific default is used.

When the recording completes, the browser adds three properties to the item's shadow variable: *duration, size* and *termchar*. These store the duration of the recording in milliseconds, the size of the recording in bytes) and the touch-tone digit used to terminate the recording, or *undefined* if the recording terminated for other reasons. So, in this example, *msg$* is the name of the shadow variable and the *msg$.duration* property stores the duration.

A recording can be played back without copying it to the server by using a *value* tag. The recording in the previous example is stored in a variable named *msg* and could be played back as follows.

```
<prompt>
    The message you recorded sounds like this:
    <value expr="msg"/>
</prompt>
```

12.3 Modal fields

A field may be defined as modal by setting the *modal* attribute to *"true"* (the default is *"false"*). This has the effect of disabling all grammars except those in the current field while the field is being interpreted.

12.4 Conditional items

The *cond* attribute may be applied to dialog items such as fields. This attribute contains an ECMAScript expression which is evaluated as a boolean result. This expression is known as a guard condition. If the condition is *false*, the field is not collected. The following form collects a dollar amount and an authorization code, the code is collected only if the amount is more than $99.

```
<form>
    <block>
        <field name="amount" type="currency">
```

```
            <prompt>
                Please say or dial the dollar
                amount now.
            </prompt>
        </field>
        <field cond="amount>99" type="number">
            <prompt>
                Please say or dial your authorization
                code now.
            </prompt>
        </field>
        <submit next="ProcessRequest.jsp"/>
    </block>
</form>
```

12.5   Root document

A VoiceXML page may optionally load a root document. This is a second page which remains loaded while the current page is loaded. It is specified by using the *application* attribute of the *vxml* tag, for example:

```
<vxml version="2.0" application="navigation.vxml">
```

It is an error for the root document itself to specify a root document, the browser will throw an error event (events are described later). It is also an error for a page to refer to a root document which cannot be loaded (because it does not exist or because there was a fetch error).

There are two main uses for the root document. First, it is a place to store global variables, in other words variables which are saved when the current page is swapped out due to a hyperlink. Second, it can be used define links, form-level grammars and other constructs which remain in scope for several different pages, perhaps even for the entire session. The root document in this way can perform roles analogous to a graphical browser's toolbar and/or a navigation bar on a graphical Web site.

Typically many or all of the pages within a single application will share the same root document.

Here is a page designed to be used as a root document.

```
<?xml version="1.0"?>
<vxml version="2.0">
   <link next="operator.vxml">
      <grammar>
         operator
      </grammar>
   </link>
   <var name="account_number"/>
</vxml>
```

This page defines a hyperlink which reacts to the word "operator" by jumping to the first dialog in *operator.vxml*. It also defines a global variable (more properly, a variable with application scope) named *account_number* which can be used to store a value which is not lost when a page is swapped out.

This means that there may be one or two pages loaded simultaneously. There are three possible scenarios for the location of the form or item currently being interpreted.

1) A stand-alone page is loaded, i.e. a page which does not define a root document; a dialog or link on that page is being interpreted.

2) A page and its root document are both loaded and a dialog or link in the page is being interpreted.

3) A page and its root document are both loaded and a dialog or link in the root document is being interpreted.

If a menu or form is defined in a root document, it will typically be given *scope="document"* attribute and will have at least one form-level grammar. This means that the dialog and its grammar(s) are always active (unless a modal field is executing). When a form-level grammar is matched, the browser starts interpreting that dialog.

12.6 Universal grammars

The browser must define exactly three universal grammars which are used to trigger the default *help*, *exit* and *cancel* events. These might be thought of as analogous to the toolbar in a graphical browser which has *Back*, *Forward*, *Home* and other buttons. The definitions of these grammars are browser-dependent and may therefore be a source of portability issues since there may be a conflict between a global grammar and an application grammar (because they might match the same user input). For maximum portability, an application can define its own versions of these grammars and disable the browser default global grammars by setting the *universals* property to *"none"*. This can be done by using a root document like the following.

```
<?xml version="1.0"?>
<vxml version="2.0">
   <property name="universals" value="none"/>
   <link event="help">
      <grammar>
         help
      </grammar>
   </link>
   <link event="cancel">
      <grammar>
         cancel
      </grammar>
   </link>
   <link event="exit">
      <grammar>
         exit
      </grammar>
   </link>
</vxml>
```

12.7 Grammar precedence

User input may match more than one grammar. If that happens, one grammar is chosen based on priority (precedence) rules. If more than one grammar with the same precedence matches, then the first grammar in source code order is chosen i.e., the grammar which appears first on the VoiceXML page. Starting with 1 as the highest priority and first to be chosen, the precedences are as follows.

1. Grammars defined in the current form item, including grammars contained in links in that field.

2. Grammars defined in the current form, including grammars contained in links in that form.

3. Grammars in links in the current page, and grammars for menus and other forms in the current page which have document scope.

4. Grammars in links in the application root document, and grammars for menus and forms in the root document which have document scope.

5. Universal grammars for default events *help*, *exit* and *cancel*.

If the current form item is modal (as defined by the *modal* attribute, which defaults to "*false*"), then all grammars except its own are turned off while waiting for input.

12.8 Menus

As we have mentioned, there are two kinds of dialogs: forms and menus. A menu is a dialog which prompts the user to make a choice, and then hyperlinks to a new dialog based on the user's response. It is specified inside a *menu* tag.

There are two main components to the menu: the prompt and the set of choices. Each choice is specified by a *choice* tag. The *choice* tag has a *next* attribute which specifies the target URL of the hyperlink to be triggered when that choice is made.

A familiar example is a touch-tone menu. The *choice* tag has an optional *dtmf* attribute which specifies the digit.

```
<menu>
   <prompt>
      For political commentary press 1,
      for stock news press 2,
      for national news press 3.
   </prompt>
   <choice dtmf="1" next="#politics"/>
   <choice dtmf="2" next="#stock_news"/>
   <choice dtmf="3" next="#national_news"/>
</menu>
```

Touch-tone digits can be automatically assigned to the choices by setting the *dtmf* attribute of the *menu* tag to *"true"*, as follows.

```
<menu dtmf="true">
   <prompt>
      For political commentary press 1,
      for stock news press 2,
      for national news press 3.
   </prompt>
   <choice next="#politics"/>
   <choice next="#stock_news"/>
   <choice next="#national_news"/>
</menu>
```

(For a touch-tone only menu, it would be recommended to disable speech input by using *<property name="inputmodes" value="dtmf"/>*. We will discuss properties later, and our menu will generally work fine without this, so we'll omit it in our examples).

We haven't gained much, because we still need to provide explicit touch-tone digits in the prompt. This process can be automated by using the *enumerate* tag. The value of the tag is a prompt which is repeated once for every choice. You could think of it as a kind of "for .. in" loop. For example,

```
<enumerate>
   For <value expr="_prompt"/>,
   please press <value expr="_dtmf"/>.
</enumerate>
```

There are two special variables which may be used within
<enumerate>. The *_prompt* variable is set to the value of (that is, the
text enclosed by) the current choice tag. The *_dtmf* variable is set to
the touch-tone digit assigned to that choice.

Putting these pieces together, we get the following.

```
<menu dtmf="true">
   <prompt>
      <enumerate>
         For <value expr="_prompt"/>,
         please press <value expr="_dtmf"/>.
      </enumerate>
   </prompt>
   <choice next="#politics">
      political commentary
   </choice>
   <choice next="#stock_news">
      stock news
   </choice>
   <choice next="#national_news">
      national news
   </choice>
</menu>
```

Now we have a full implementation of a touch-tone menu where
the browser automatically assigns a digit to each choice and
constructs an appropriate prompt.

If the *enumerate* tag is empty, the browser will substitute a default.
The value of that prompt is browser-specific.

The *choice* tag may contain anything that is valid inside a *prompt*
tag.

Menus can also use grammars to define the input. For example,

```
<menu>
   <prompt>
      Please choose political commentary,
      stock news or national news now.
   </prompt>
```

```
<choice next="#politics">
   <grammar>
      political commentary
   </grammar>
</choice>

<choice next="#stock_news">
   <grammar>
      stock news
   </grammar>
</choice>

<choice next="#national_news">
   <grammar>
      national news
   </grammar>
</choice>
</menu>
```

The hyperlink will be triggered in the first choice where the user input matches its grammar.

As with touch-tone digits, the menu can automatically generate the grammars to be matched. By default, the generated grammar is an exact match to the text value of the choice. Consider this revised example.

```
<menu>
   <prompt>
      Please choose political commentary,
      stock news or national news now.
   </prompt>

   <choice next="#politics">
      political commentary
   </choice>

   <choice next="#stock_news">
      stock news
   </choice>
```

```
    <choice next="#national_news">
       national news
    </choice>
</menu>
```

The generated grammars would be exactly the same as the explicit grammars given in the previous example.

We can again use the *enumerate* tag to automate the menu prompt. Remember that the _prompt variable contains the text value of the *choice* tag.

```
<menu>
    <prompt>
       Please choose one of the following:
       <enumerate>
          <value expr="_prompt"/>,
       </enumerate>
       now.
    </prompt>

    <choice next="#politics">
       political commentary
    </choice>

    <choice next="#stock_news">
       stock news
    </choice>

    <choice next="#national_news">
       national news
    </choice>
</menu>
```

There are two types of automatic grammar for the choices, specified by the *accept* attribute of the *choice* tag. The default is *accept="exact"*, which requires a complete match to the phrase. The alternative is *accept="approximate"*, which matches any subphrase. In our example, we need an exact match to distinguish "national news" from "stock news", but we could also accept "political" or "commentary" alone for the first choice, so it would make sense to use:

```
<choice accept="approximate">
  political commentary
</choice>
```

The exact meaning of "matches any sub-phrase" will be browser-dependent. The typical interpretation is to match any set of words from the phase occurring in the same order that they are specified in the phrase, i.e. for a phrase *word1 word2 word3...*, match:

$$word1_{opt} \; word2_{opt} \; word3_{opt}...$$

The text used for automatic grammar generation and *<enumerate>* is built from all text found in the complete sub-tree under the *choice* tag. For example,

```
<choice accept="approximate">
  <audio src="PolCom.wav">
    political commentary
  </audio>
</choice>
```

The *menu* tag has an optional attribute called *scope*. The scope of a menu is dialog or document. By default, the menu has dialog scope, which means that its choices are active only when the menu is active. If the menu is given *scope="document"*, then the menu choices are active the whole time the page is loaded, in which case when inactive it is like a document-level set of hyperlinks, one link for each choice.

12.9   Field options

When a field needs to collect a simple choice like "red", "green" or "blue", then the *option* tag provides a convenient alternative to creating a grammar. It functions in a similar way to a menu *<choice>*. The following example illustrates how it is used.

```
<form>
   <field name="color">
      <prompt>
         Please select a color.
         <enumerate>
            Say <value expr="_prompt"/> or
            press <value expr="_dtmf"/>.
         </enumerate>
      </prompt>
      <option dtmf="1">red</option>
      <option dtmf="2">green</option>
      <option dtmf="3">blue</option>
   </field>
</form>
```

The *dtmf* attribute specifies the touch-tone for that choice. If *dtmf* is not specified, the browser will assign the next free digit automatically, unlike <choice> you can't prevent touch-tone digits from being assigned.

The *value* attribute specifies the value to assign to the field variable if this option is chosen. By default, this is the value of the tag, i.e. the text contained within the tag. As this example shows, the *enumerate* tag supports <option> in a very similar way to <choice>.

You can think of <option> as similar to specifying a value in a pull-down list or a set of radio buttons displayed by a form in a graphical browser, in the same way that the *boolean* built-in type is analogous to a check-box.

12.10 Mixed-initiative forms

The forms we have seen so far have been machine-directed, meaning that the order in which the fields are collected is determined by the VoiceXML page. A more flexible kind of conversation can be achieved using a mixed-initiative dialog in which the user can fill in multiple fields by speaking a single phrase. If any fields remain unfilled, the browser proceeds to direct the dialog one field at a time to complete the form in the same way we have seen in earlier examples.

The key to creating a mixed-initiative form is to create a grammar which can fill in more than one field. This requires so-called semantic markup of the grammar, where parts of the grammar are flagged as setting a particular field to a particular value. How this will work has not yet been decided at the time of writing. We are going to suppose that there is a *field* attribute of the grammar tags which names the field to be filled in by that part of the grammar. You will need to consult your browser documentation or a more recent version of the VoiceXML standard to find out how this really works.

We'll create a mixed-initiative form which accepts the nine different phrases "I would like to buy a *[red/green/blue] [car/truck/minivan]*" and sets two fields, *color* and *vehicle_type*. First we need a grammar.

```
<grammar>
    I would like to buy a
    <one-of field="color">
        <item>red</item>
        <item>green</item>
        <item>blue</item>
    </one-of>
    <one-of field="vehicle_type">
        <item>car</item>
        <item>truck</item>
        <item>minivan</item>
    </one-of>
</grammar>
```

We'll call this file *"cars.gram"*.

In a mixed-initiative form, the grammar is associated with the form (i.e., is a child tag of the form) rather than with any particular field.

A mixed-initiative form should begin by playing a prompt which encourages the user to speak a phrase which matches the form-level grammar. This is done by using an *initial* tag. This is a special type of form item. It will typically contain *<prompt>* and/or *<audio>* tags to play the prompt. It may also use *<enumerate>* and

<value> to build all or part of the prompt automatically from
menu choices or by calling script code, respectively. Note that you
can't have prompts directly under the *form* tag (unlike *<menu>*,
which does allow you to do this).

The form could look like this.

```
<form>
    <grammar src="cars.gram"/>
    <initial>
        <prompt>
            What kind of vehicle would you like?
        </prompt>
    </initial>
    <field name="color"/>
    <field name="vehicle_type/>
</form>
```

This form has a serious flaw. If the user doesn't fill in all the fields
by matching *cars.gram,* then the browser doesn't have a way to fall
back to machine-directed processing (because there are no field-
level prompts or grammars).

To improve on this, we need to provide prompts and grammars
for the individual fields. We'll start by upgrading our grammar to
give named rules for the color and vehicle type.

```
<grammar>
    I would like to buy a
    <ruleref name="color_rule"/>
    <ruleref name="vehicle_type_rule"/>
    <rule name="color_rule" scope="public">
        <one-of field="color">
            <item>red</item>
            <item>green</item>
            <item>blue</item>
        </one-of>
    </rule>
```

```
        <rule name="vehicle_type_rule" scope="public">
            <one-of field="vehicle_type">
                <item>car</item>
                <item>truck</item>
                <item>minivan</item>
            </one-of>
        </rule>
    </grammar>
```

We can now upgrade our form as follows.

```
<form>
    <grammar src="cars.gram"/>
    <initial>
        <prompt>
            What kind of vehicle would you like?
        </prompt>
    </initial>
    <field name="color">
        <prompt>
            Please choose a color for your vehicle
            by saying red, green or blue.
        </prompt>
        <grammar src="cars.gram#color_rule"/>
    </field>
    <field name="vehicle_type>
        <prompt>
            Please choose a type of vehicle by
            saying car, truck or minivan.
        </prompt>
        <grammar src="cars.gram#vehicle_type_rule"/>
    </field>
</form>
```

With this form, the conversation might proceed as follows.

BROWSER *Starts with <initial>:*	What kind of vehicle would you like?
USER:	An orange sports car.
BROWSER *Triggers nomatch event, no form-level handler is provided so browser default is used:*	I'm sorry, I didn't understand what you just said.
BROWSER *The <initial> item has already been visited, so it is already filled; visit first unfilled field, which is "color":*	Please choose a color for your vehicle by saying red, green or blue.
USER:	Red.
BROWSER: *Visits first unfilled field, which is now "vehicle_type":*	Please choose a type of vehicle by saying car, truck or minivan.
USER:	Truck.

The browser first processes ("visits") the *<initial>* item. It is considered to be filled in when it has been visited once. When the user input has been processed, the browser will proceed to visit any unfilled fields in the order they appear (this is the standard form processing algorithm).

12.11 Subdialogs

A subdialog is a form which is called from another form and returns, like a subroutine. Subdialogs are useful for breaking up complex forms into simpler pieces and for re-using code.

As an example, we'll construct a form which has two fields: credit card number and expiration date. The expiration date will be collected by using a subdialog.

The *return* tag is used to return from a subdialog to the calling form. The *namelist* attribute is used to give a list of field names whose values are transmitted back to the calling form. Unlike the *submit* tag, the default is an empty namelist, not all fields.

Here is the form we'll call as a subdialog.

```
<form name="get_exp">
   <field name="exp" type="date">
      <prompt>
         Please say the expiration date.
      </prompt>
      <filled>
         <return namelist="exp"/>
      </filled>
   </field>
</form>
```

A subdialog is called by using the *subdialog* tag. It is a dialog item which is initially unfilled, like all dialog items. When it is processed, the subdialog is called, the dialog to visit is specified by the *uri* attribute, which gives the URL.

Here is a form that invokes *get_exp* as a subdialog.

```
<form>
    <field name="cardnr" type="digits">
        <prompt>
            Please say or dial your card number now.
        </prompt>
    </field>
    <subdialog name="expdate" uri="#get_exp"/>
    <block>
        <submit uri="Validate.jsp"/>
    </block>
</form>
```

The *return* tag returns values by setting properties in the variable for the calling *subdialog* item. Each variable in the *return* tag's namelist sets one property in the variable. In this example, the namelist contains one variable called *"exp"* and the calling *<subdialog>* is named *"expdate"*, so one property named *expdate.exp* is set. The *submit* command assumes a namelist with all variables, which in this case is equivalent to *namelist="cardnr exp"* which would send something like:

```
Validate.jsp?ccnr=0123456789&expdate.exp=200306??
```

If an ECMAScript variable in the namelist has properties, as in this case, each property is sent individually. Remember that date fields which are not set by user input are designated by ?s, we imagine here that the user said "June 2003".

12.12 Events

We have already encountered two events: *nomatch* and *noinput*, which are handled by *<nomatch>* and *<noinput>*, respectively. To refresh your memory, here is a field with a *noinput* handler.

```
<field name="cardnr" type="digits">
    <noinput>
        I'm sorry, I didn't hear anything.
    </noinput>
    <prompt>
        Please say or dial your card number now.
    </prompt>
</field>
```

An event is a message or notification of a particular state. Events are generally things which happen to the application at unpredictable times, such as the user hanging up the call in the middle of a dialog. When a condition happens, the corresponding event is said to be triggered, raised or thrown. When an event is triggered, the browser interrupts its current processing and searches for code which can accept the event. A piece of code which handles an event is known as an event handler and is said to catch it.

There is a similar concept in Java and C known as an exception. In Visual Basic exceptions are handled by using *On Error Go To*.

The *noinput* and *nomatch* tags are a convenient shorthand for common forms of the *catch* tag. The *event* attribute gives the name of the event, so the previous example is exactly equivalent to the following.

```
<field name="cardnr" type="digits">
   <catch event="noinput">
      I'm sorry, I didn't hear anything.
   </catch>
   <prompt>
      Please say or dial your card number now.
   </prompt>
</field>
```

As with several other constructs in VoiceXML, an event handler can have one of the four scopes: application (i.e., root document, which wraps the current page), document (current page), dialog (current form or menu), or item (the current *<field>*, *<block>*, *<initial>* etc.). The scope is determined by the parent tag, so if for example the event handler is a child of the *form* tag for the current dialog, then the handler has dialog scope.

When an event is thrown, the browser searches outwards from the current scope until an event handler is found. The first handler found is the one which is executed, if there are also handlers in outer scopes these are ignored. So, for example, if a *noinput* event is thrown inside a field, the browser will look for event handlers in the following order: current field, current form, current page and finally in the root document (if there is one). If no handler is found, a default handler supplied by the browser will be executed. This handler can be thought of as having session scope.

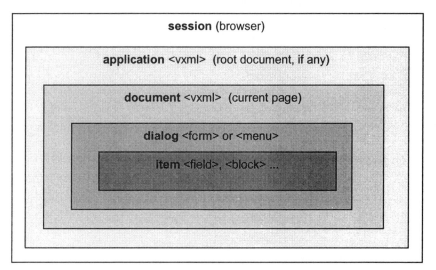

VoiceXML scopes.

When the catch element completes, by default execution will continue in the dialog which threw the event, in which case interpretation continues in the usual way at the first unfilled field. The catch block may contain a hyperlink such as *goto* or *submit*, which transfers to a new dialog.

The *throw* tag can be used to throw an event. This can be used for example to send an event that further up the scope chain, perhaps to an event handler in the root document or the browser's default handler. For example,

```
<field name="cardnr" type="digits">
   <noinput>
      <assign name="noinput_count"
        expr="noinput_count+1"/>
      <throw event="noinput"/>
   </noinput >
```

```
    <prompt>
        Please say or dial your card number now.
    </prompt>
</field>
```

You can invent your own events. It is recommended that customer error events be named *error.* followed by a reverse domain name, such as *error.com.dialogic.too_many.*

```
<-- In the root document: -->
<catch event="error.com.dialogic.too_many "/>
    <prompt>
        Sorry, you are obviously not smart enough
        to use our system. Good bye.
    </prompt>
    <disconnect/>
</catch>

<-- In the current page: -->
<field name="cardnr" type="digits">
    <noinput>
        <assign name="error_count"
          expr="error_count+1"/>
        <if cond="error_count>9">
          <throw
            event="error.com.dialogic.too_many "/>
        </if>
        <throw event="noinput"/>
    </noinput >
    <prompt>
        Please say or dial your card number now.
    </prompt>
</field>
```

Events which are named with a period, such as *com.dialogic.too_many* or *error.badfetch*, can be caught by handlers with a name which matches up to one of the periods. So, for example, *com.dialogic* matches *com.dialogic.to_many*, and *error* matches *error.badfetch*. Within a given scope, the first handler which matches will be executed, even if there is a handler with a more complete name in the same scope or a higher scope. Suppose a page has the following event handlers.

(1) *<catch event="error">* with item scope.
(2) *<catch event="error.badfetch">* later in the item scope.
(3) *<catch event="error.badfetch.http.404">* with form scope.

If an *error.badfetch.http.404* event is thrown in field scope, then handler (1) will be executed because it is the best match within the current scope, even though there is a better match in the same scope. In fact, handler (2) will never be executed since (1) will always match first.

The browser will catch all events which are not caught by the application. The action taken is shown in the following table.

Event	Default handler
cancel	Do nothing (continue to interpret the current dialog).
error	Message, *<exit>*.
exit	*<exit>*.
help	Message, *<reprompt>*.
noinput	*<reprompt>*.
nomatch	Message, *<reprompt>*.
telephone. disconnect	*<exit>*.
Other	Message, *<exit>*.

"Message" indicates that the browser plays a message, the content of the message is browser-dependent.

The tags *noinput, nomatch, error, cancel* and *help* are convenient shorthands for commonly used events, for example *<help>* is equivalent to *<catch event="help">*.

12.13 Properties

A property is a named value which determines a browser setting. For example, the property *inputmodes* determines whether speech, touch-tones or both is enabled for collecting user input. Allowed values are *"dtmf"*, *"voice"*, *"speech"* (the last two are synonyms, they have exactly the same effect), or a whitespace-separated list such as *"dtmf voice"*. For example,

```
<property name="inputmodes" value="dtmf"/>
```

Properties can be specified at application (i.e., root document), document (i.e., the current page), dialog (form or menu) or form item level. Property values are inherited by child tags unless overridden by a different value. Here is our earlier menu example with speech input disabled (since it only accepts touch-tones).

```
<menu>
    <property name="inputmodes" value="dtmf"/>
    <prompt>
       For political commentary press 1,
       for stock news press 2,
       for national news press 3.
    </prompt>
    <choice dtmf="1" next="#politics"/>
    <choice dtmf="2" next="#stock_news"/>
    <choice dtmf="3" next="#national_news"/>
</menu>
```

There may be browser-specific properties. To minimize collisions it is recommended that such properties be named like Java packages by using a reverse domain name, for example *com.dialogic.positive_voice_detection*. If a browser encounters an unsupported property, it must ignore it (the *error.unsupported.property* event will not be thrown).

12.14 Browser extensions

VoiceXML defines the *object* tag to allow browser-specific extensions. Vendors should use <*object*> to define language extensions rather than adding proprietary tags.

The *object* tag is a form item, so it a child of a *<form>* and is visited like fields, blocks and other items when the form is interpreted. Like other items, it is considered unfilled when the form is initialized and like *<initial>* it is considered filled when it has been visited once.

The *object* tag has several optional attributes, including *classid* and *data*, which specify which object to use. The interpretation is browser-dependent.

As an example, we'll imagine that our browser supports loading a Java class and invoking a method on that class. The *classid* attribute specifies the method to call, the *data* attribute holds the URL of the class file. The following example would call a Java method.

```
<object data="billing.class" classid="FreeCall"/>
```

Invoking the object can set its variable. Let's suppose that calling the *Billing.GetRates* method sets the variable to the per minute rate for the call. The following form gets the rate then speaks it to the user.

```
<form>
    <object name="rate" data="billing.class"
      classid="GetRates"/>
    <block>
        You will be charged <value expr="rate"
          class="currency"/> per minute.
    </block>
</form>
```

Multiple values could be returned by setting properties of the variable.

As with other dialog items, this object's variable can be cleared (set to unfilled) by using *<clear name="rate ">* or by setting *rate* to *undefined* in a script expression, which would cause the object item to be re-visited.

If an *object* tag refers to an unsupported object type, the browser will throw an *error.unsupported.object* event. Some browsers may support no object types at all.

12.15 Passing parameters to subdialogs and objects

The *param* tag is similar to *<assign>*. It is used to set the value of a variable which is passed to the called subdialog or object. It is similar to setting the value of a subroutine parameter. The *param* tag is a child of *<subdialog>* or *<object>*.

The *name* attribute specifies the name of the variable to set, the *value* or *expr* attribute is used to set the value. The expression is evaluated in the scope of the calling form, in the case of a subdialog the resulting variable is set in the scope of the called dialog and must be declared within that dialog by using *<var>*.

We'll expand on our earlier example of a subdialog which gets the expiration date so that the main form gets the card type (Visa or Mastercard) and the subdialog includes the card type in the prompt (e.g., "Please say the expiration date of your Visa now"). We declare a variable *cardtype* in the subdialog which will receive the card type.

```
<-- This form is called as a subdialog -->
<form name="get_exp">
   <var name="cardtype"/>
   <field name="exp" type="date">
      <prompt>
         Please say the expiration date
         of your <value expr="cardtype"/> now.
      </prompt>
      <filled>
         <return namelist="exp"/>
      </filled>
   </field>
</form>
```

```
<-- Main form which calls the subdialog -->
<form>
    <field name="vendor">
        <grammar>
            visa | mastercard
        </grammar>
        <prompt>
            We accept only Visa and Mastercard.
            Please say your card type now.
        </prompt>
    </field>
    <field name="cardnr" type="digits">
        <prompt>
            Please say or dial your card number now.
        </prompt>
    </field>
    <subdialog name="expdate" uri="#get_exp">
        <param name="cardtype" expr="vendor"/>
    </subdialog>
    <block>
        <submit uri="Validate.jsp"/>
    </block>
</form>
```

12.16 Transferring calls

The *transfer* tag is a form item which requests the browser to transfer a call. The typical use envisaged for this tag is to transfer a call to a live operator or agent.

There are two types of transfer: bridged and unbridged. In both cases, the browser initiates a second call and connects the party reached by this second call ("the called party") to the user.

A bridged transfer creates a three-way call where the browser remains on the call together with the user and the called party.

In an unbridged transfer, the browser disconnects from the call as soon as the second call has been answered, leaving the user in a one-on-one conversation with the called party. The browser throws a *telephone.disconnect.transfer* event when it disconnects itself.

The *transfer* tag defines a dialog item which is visited in the same way as other dialog items. The *dest* attribute specifies the URL of the number to dial, the *phone://* scheme is used for conventional phone numbers. The *bridge* attribute is set to "*false*" for an unbridged transfer (the default) or "*true*" for a bridged transfer. The *connecttimeout* attribute is set to the maximum time to wait for a connect.

During the call progress analysis phase where the browser is waiting for the second call to be answered, user input is monitored and if it matches a grammar defined in the *transfer* tag, then the transfer will be aborted. The *transfer* tag is modal, meaning that grammars outside the tag are always inactive while the tag is being visited, even if those grammars are in scope.

When the call progress phase is completed, the variable associated with the *transfer* item is set to the call progress result and any *<filled>* sub-tags are executed. Possible values for the result are shown in the following table.

Result	Description
busy	The called number was busy.
far_end_disconnect	Call was completed but called party hung up.
near_end_disconnect	Call was completed but user hung up.
network_busy	Called number could not be reached due to network congestion (in the US, indicated by a "fast busy" tone).
network_disconnect	Call was completed but terminated by the network.
noanswer	Any other result.

The following form attempts a supervised transfer and plays an appropriate message if the call was not completed.

```
<form>
    <transfer name="xfer" dest="phone://8005551212"
      connecttimeout="30s" bridge="false">
        <prompt>
            Transferring your call now.
        </prompt>
        <filled>
            <if cond="xfer='busy'">
                <prompt>
                    The number was busy, please
                    try again later.
                </prompt>
            <else/>
                <prompt>
                    Sorry, could not reach the number.
                </prompt>
            </if>
        </filled>
    </transfer>
</form>
```

12.17 Tracking a call

Applications sometimes need to track an individual call from page to page, or even from endpoint to endpoint in the case where a call is transferred or conferenced. There a few technical issues to consider: one browser is likely to be used by many different callers, one Web server may serve several different VoiceXML browsers, and the underlying HTTP protocol does not support session tracking.

With graphical browsers, cookies can track a user's session. In VoiceXML, cookies are much less useful since a browser can store only one cookie per site and there may be many simultaneous calls for that site.

The ANI digits (the phone number of the caller's line) can play a role similar to a cookie since in many cases they uniquely identify a user, at least for the duration of a given call (it should not be possible for two callers to have the same ANI digits at the same time). The *session.telephone.ani* variable contains'the ANI digits are can be submitted to the server via a *<submit>* tag when the server needs to identify the call.

If ANI digits are not available, a call identification number (CIN) can be stored in the root document. When a request for the start page is generated, a server script assigns a new CIN an embeds an assignment to the root document CIN variable in the page which is sent to the browser. Here is a root document which declares the CIN variable, let's suppose it is found in file name *app_root.vxml*.

```
<?xml version="1.0"?>
<vxml version="2.0">
   <var name="CIN"/>
</vxml>
```

The start page for the application might look like the following, a new version of the page with a different CIN would be generated for each call by using a server script.

```
<?xml version="1.0"?>
<vxml version="2.0" application="app_root.vxml">
   <script>
       application.CIN = 01234567890;
   </script>
   <form>
      <block>
         <goto next="app_home.vxml"/>
      </block>
   </form>
</vxml>
```

Now *application.CIN* can be used like *session.telephone.ani*.

12.18 Executable content

Now that we have covered most of the VoiceXML language, we can return to the term "executable content" and give it a more precise definition.

Executable content is part of a VoiceXML document which behaves like a conventional scripting language: it is interpreted in the order it appears except as modified by tags such as *<if>* and *<goto>*. The following table gives a complete list of legal executable content tags.

Tag	Description
`<assign>`	Assign to a variable.
`<audio>`	Play pre-recorded or streaming audio.
`<clear>`	Set one or more variables to *undefined*, by default all the variables in the current form are cleared.
`<disconnect>`	Hang up and terminate browser session.
`<enumerate>`	Output contents as for *<prompt>* once for each item in the current form.
`<exit>`	Terminate browser session, subsequent actions if any are browser-dependent.
`<goto>`	Terminate current dialog and start interpreting a new dialog.
`<if>`	Conditionally execute. Value of *<if>* may be any valid executable content plus *<else>* and *<elseif>*.
`<log>`	Output value to debug log.
`<prompt>`	Output using text-to-speech. The *prompt* tag may contain any SSML tags, including *<audio>*.
`<reprompt>`	Set flag indicating that the browser should re-prompt for the current dialog item.
`<return>`	Return from subdialog.
`<script>`	Execute ECMAScript code.
`<submit>`	Send form item variable values and request new page from server.

Tag	Description
`<throw>`	Throw event.
`<value>`	Insert expression into text to be spoken.
`<var>`	Declare a variable.

It is also legal to use text as executable content, in which case it is output using text-to-speech. As previously noted, we do not recommend doing this because such text cannot be marked up, we recommend always using *<prompt>*.

There are several tags which contain executable content (and may not containing anything else). They are listed in the following table.

Tag	Description
`<block>`	A form item which is a placeholder for executable content. The content is executed when the item is visited.
`<catch>`	Event handler. The shorthand equivalents *<error>*, *<help>*, *<noinput>* and *<nomatch>* also contain executable content.
`<filled>`	Found inside a form item (or form). Is executed when the form item (or given set of form items) has been filled by user input.

There is one special case. The root tag *<vxml>* may contain *script* tags, which are executed when the page loads, but no other executable content tags.

12.19 Dialog items

As we have seen, a VoiceXML form is primarily composed of items, which by default are visited one by one in order until they are filled. An item is considered "filled" when the variable associated with the item has a value other than *undefined*. An item variable can be assigned *undefined* by using a *script* tag or *expr* attribute or by using the *clear* tag, this would cause the item to be re-visited.

All form items have the following attributes, some may have other attributes.

Form item attribute	Description
name	(Required). Name of the ECMAScript variable for this item.
expr	(Optional). Initial value for the item variable.
cond	(Optional). Guard condition. Item is not visited if condition is *false*.

Form item tags are summarized in the following table.

Tag	Description
<field>	Collects user input which is assigned to its item variable.
<block>	Run executable content.
<initial>	Run executable content and activate form-level grammars. For collecting user input which can set more than one field.
<subdialog>	Call another dialog and return.
<object>	Execute a browser-specific language extension.

`<record>`	Record audio, the audio is assigned to the form item variable.
`<transfer>`	Transfer call, if a bridged transfer then save result in item variable.

12.20 Resource fetching

Tags which require a resource to be fetched from another URL have a common set of attributes, as shown in the following table.

Attribute	Allowed values
`caching`	`"safe"` `"fast"`
`fetchhint`	`"safe"` `"prefetch"` `"stream"`
`fetchtimeout`	Time value, number followed by "s" for seconds or "ms" for milliseconds, e.g. "2.5s".
`fetchaudio`	(Supported only by those tags which fetch VoiceXML pages). URL of an "hourglass" sound to play while fetching.

The *caching* property specifies whether the browser is permitted to cache the resource locally. If *caching="safe"*, the browser must fetch a new copy of the resource even if it has not expired according to the time stamp on the HTTP response header. If *caching="fast"*, the browser may, but is not obligated to, keep a local copy of the resource and is not obligated to fetch the resource again until a) it is needed and b) the cache has expired according to the HTTP header. The default value of *caching* is *"fast"*.

The value of *fetchhint* specifies in more detail when the resource may be fetched. If set to *"safe"*, the resource may not be fetched until it is needed. If set to *"prefetch"*, it may be fetched at the time the page containing the URL is loaded. If set to *"stream"*, this indicates that the browser should begin output without waiting for the complete resource to be fetched. The default value of *fetchhint* is not defined by the current VoiceXML specification, so is defined by the browser.

The *fetchaudio* property specifies the URL of a sound to be played while fetching a VoiceXML page. You can think of it as being like an "hourglass" sound, an appropriate choice might be the sound of a ticking clock. The play is interrupted when the fetch is completed.

Default values for all these attributes can be specified by setting properties. For example,

```
<property name="caching" value="fast"/>
<property name="fetchhint" value="prefetch"/>
<property name="fetchtime" value="2500ms"/>
<property name="fetchaudio" value="ticking.wav"/>
```

These properties can be provided in item, dialog, document or application scope. The innermost scope determines the value. In other words, if a property is defined in dialog scope, this overrides the property specified in document or application scope.

If the *fetchaudio* attribute is not specified and no *fetchhint* property has been set, a compliant browser will be silent while fetching, there should be no browser-dependent default sound.

12.21 The Form Interpretation Algorithm

At the core of every VoiceXML browser is the Form Interpretation Algorithm (FIA). The algorithm is defined by the VoiceXML language specification and specifies the actions the browser should take when interpreting a page. Note that the following description is deliberately simplified to make it easier to understand. A more complete description can be found in the official VoiceXML specification.

In broad outline, the algorithm looks like this. We assume we have a URL of a page to execute. When the browser starts, the URL will be supplied to it somehow, for example by mapping the DNIS digits to a URL in a configuration database. Once this page is loaded, new URLs can be provided through hyperlinks in *<goto>*, *<submit>* and *<link>* tags.

1. Fetch the new page.

2. Initialize the page by declaring variables found in *var* tags at the top level, running any *script* tags at the top level and activating grammars for all *link* tags which have document scope.

3. If the URL has an anchor (also called fragment, that is a URL ending in ...#*name*), then find the dialog with that name. Otherwise find the first dialog in the page.

4. Initialize the new dialog by activating <*link*> grammars in this dialog which have dialog scope, setting all form item variables to *undefined*, declaring variables found in *var* tags.

5. Find the first form item which is not marked as filled, in other words, which has a form item variable with value *undefined*, and execute that item. If there is no unfilled form item, terminate the browser session in the same way as <*exit*>.

6. If a hyperlink has been triggered (as a result of user input matching a <*link*> grammar or because a <*goto*> or <*submit*> was encountered in executable code), go to step 1 (if the form is on a new page) or step 4 otherwise.

7. If no hyperlink was triggered, go back to step 5.

Step 5 describes finding and executing a dialog item. This proceeds in three phases: select, collect and process.

12.21.1 Select phase

The goal of the select phase is to identify the next form item to visit. This proceeds as follows.

5a. If this point was reached by <*goto*> with the *nextitem* attribute set to indicate an attribute name, then use the specified item. Otherwise proceed to step 5b.

5b. Find the first item where the guard condition specified by the *cond* attribute (if any) is not *false* and where the form item variable is set to *unfilled*.

12.21.2 Collect phase

The goal of the collect phase is to prompt for user input and match user input to a grammar.

The form item chosen in the select phase is visited, which performs actions that depend on the type of form item. If a field item or *initial* tag is visited, the browser plays prompts based on the field item's prompt counter and the prompt conditions. Then it waits for user input and waits for a grammar recognition or for an event. For a field item, item-level grammars will be active. For an *initial* tag, form-level grammars will be active.

A *block* tag is visited by setting its form item variable to true, running the executable content of the tag. No input is collected, and the process phase which would otherwise come next is skipped.

12.21.3 Process phase

The process phase determines the action to take based on the user input or event obtained during the collect phase. The browser is guaranteed to get one or the other because a *noinput* event will be triggered if nothing else happens.

If the user input matches a grammar, one or more variables are set as appropriate (a field-level grammar sets the variable for its field, a form-level grammar may set more than one field).

If the user input matches a hyperlink, the link is executed, which has the effect of terminating the current form and moving to step 1 (new page and new dialog) or step 4 (new dialog on current page).

If an event was thrown, the browser searches for and executes the appropriate event handler.

13 VoiceXML Reference: Tags

<assign>
Assign a value to a variable.

Parent Tags

<block> <catch> <error> <filled> <help> <if> <noinput> <nomatch>

Content Syntax

The value of this tag must be empty.

Attributes

Name	Req/opt	Default	Valid values
expr	Required		Text
name	Required		Text

Compatibility

VoiceXML 1.0 Yes, 2.0 Yes.

Description

Assigns a value to an ECMAScript variable. The name of the variable is specified by the name attribute, the ECMAScript expression in the expr attribute is evaluated and assigned to that variable.

The variable could be a dialog item variable (defined by the name attribute of <field> etc.), or a variable declared by <var>. The variable must be declared before being used and must be in scope. A variable from an outer scope can be referenced by using scopename.varname, where scopename is application, document or dialog. This is required in the case where the variable in the outer scope has been obscured by a variable of the same name, is optional otherwise.

\<audio\>
Play audio data.

Parent Tags
\<audio\> \<block\> \<catch\> \<choice\> \<emphasis\> \<enumerate\> \<error\>
\<field\> \<filled\> \<help\> \<if\> \<initial\> \<menu\> \<noinput\> \<nomatch\>
\<object\> \<p\> \<paragraph\> \<prompt\> \<prosody\> \<record\> \<s\>
\<sentence\> \<subdialog\> \<transfer\> \<voice\>

Content Syntax
The value of this tag may be empty, or may contain one or more
elements, where an element is either parsed text or one of the following
tags.

\<audio\> \<break\> \<emphasis\> \<enumerate\> \<mark\> \<p\> \<paragraph\>
\<phoneme\> \<prosody\> \<s\> \<say-as\> \<sentence\> \<value\> \<voice\>

Attributes

Name	Req/opt	Default	Valid values
caching	Optional		"fast" "safe"
expr	Optional		Text
fetchhint	Optional		"prefetch" "safe" "stream"
fetchtimeout	Optional		Text
src	Optional		Text

Compatibility
VoiceXML 1.0 Yes, 2.0 Yes.

Description
Plays pre-recorded or streaming audio. The URL of the source must be
specified either by the src attribute, which specifies a fixed URL, or by
the expr attribute, which specifies an ECMAScript expression which is
evaluated to determine the URL.

For streaming audio, you should set caching="safe" and
fetchhint="stream".

The value of the tag (the text between the begin and end tags) specifies "alternative content", i.e. text to be played by speech synthesis in the event that the audio cannot be fetched from the given URL due to a server problem, network overload or other issues. If alternative content is played, it is played exactly as if it had been enclosed in a <prompt> tag.

\<block\>

Item which executes code.

Parent Tags

\<form\>

Content Syntax

The value of this tag may be empty, or may contain one or more elements, where an element is either parsed text or one of the following tags.

\<assign\> \<audio\> \<clear\> \<disconnect\> \<enumerate\> \<exit\> \<goto\> \<if\> \<log\> \<prompt\> \<reprompt\> \<return\> \<script\> \<submit\> \<throw\> \<value\> \<var\>

Attributes

Name	Req/opt	Default	Valid values
cond	Optional		Text
expr	Optional		Text
name	Optional		Text

Compatibility

VoiceXML 1.0 Yes, 2.0 Yes.

Description

A dialog item which contains executable content. When the block is visited, the value of the tag (the content between the begin and end tag) is executed.

<break>
Insert a pause.

Parent Tags

<audio> <choice> <emphasis> <enumerate> <p> <paragraph>
<prompt> <prosody> <s> <sentence> <voice>

Content Syntax

The value of this tag must be empty.

Attributes

Name	Req/opt	Default	Valid values
size	Optional	"medium"	"large" "medium" "none" "small"
time	Optional		Text

Compatibility

VoiceXML 1.0 **Yes**, 2.0 **Yes**.

Description

Inserts a period of silence into synthesized speech.

<catch>

Catch an event.

Parent Tags

<field> <form> <initial> <menu> <object> <record> <subdialog>
<transfer> <vxml>

Content Syntax

The value of this tag may be empty, or may contain one or more elements, where an element is either parsed text or one of the following tags.

<assign> <audio> <clear> <disconnect> <enumerate> <exit> <goto> <if>
<log> <prompt> <reprompt> <return> <script> <submit> <throw>
<value> <var>

Attributes

Name	Req/opt	Default	Valid values
cond	Optional		Text
count	Optional		Text
event	Required		Text

Compatibility

VoiceXML 1.0 Yes, 2.0 Yes.

Description

Defines an event handler. The event to be caught is specified by the name attribute. The count and cond attributes can be used to control when a handler is considered eligible to catch an exception.

The event handler contains executable content, this content is executed when the handler catches an event.

When an event is thrown, the browser searches the current scope for event handlers in the order in which they are written in the page. The name of the event matches if the name handler matches the name attribute up to a period. Exactly one handler will catch a given event. So, if an event a.b.c is thrown, and the event handler name is a.b, the name matches. The first matching event handler in a given scope is executed, even if there is a closer match later in the same scope, so a handler for a.b will be executed even if there is a later handler for a.b.c.

If no eligible handler with a matching name is found in the current scope, the browser searches outwards for a scope with a matching handler.

\<choice\>
Defines a menu option.

Parent Tags

\<menu\>

Content Syntax

The value of this tag may be empty, or may contain one or more elements, where an element is either parsed text or one of the following tags.

\<audio\> \<break\> \<emphasis\> \<enumerate\> \<grammar\> \<mark\> \<p\>
\<paragraph\> \<phoneme\> \<prosody\> \<s\> \<say-as\> \<sentence\> \<value\>
\<voice\>

Attributes

Name	Req/opt	Default	Valid values
accept	Optional	"exact"	"approximate" "exact"
caching	Optional		"fast" "safe"
dtmf	Optional		*Text*
event˙	Optional		*Text*
expr	Optional		*Text*
fetchaudio	Optional		*Text*
fetchhint	Optional		"prefetch" "safe" "stream"
fetchtimeout	Optional		*Text*
next	Optional		*Text*

Compatibility

VoiceXML 1.0 Yes, 2.0 Yes.

Description

Specifies one choice in a \<menu\>.

When the choice is selected by the user by making appropriate input, the choice tag will either transition to a new dialog or throw an event. To indicate the action to take, exactly one of the next, expr or event attribute must be set. The next and expr attributes specify the URL of the next dialog.

A choice may contain a prompt to be spoken to explain the choice to the user. This prompt is available through the special _prompt variable in the <enumerate> tag.

The choice may contain one or more grammars which specify the user input which selects it. Alternatively, the dtmf attribute can be set to indicate a touch-tone digit which selects it. If the parent <menu> tag has dtmf="true", then a digit will be assigned automatically to this choice if one is not specified. The digit assigned to the choice is available through the special _dtmf variable in the <enumerate> tag.

<clear>
Re-set form items.

Parent Tags
<block> <catch> <error> <filled> <help> <if> <noinput> <nomatch>

Content Syntax
The value of this tag must be empty.

Attributes

Name	Req/opt	Default	Valid values
namelist	Optional		Text

Compatibility
VoiceXML 1.0 Yes, 2.0 Yes.

Description
Re-sets one or more form items to its initial state. This means that its variable is set to undefined and its prompt and event counters are set to zero. By default, all items in the current form are cleared. A specific list of items to clear can be specified by using the namelist attribute.

Clearing an item makes it eligible to be visited again providing that its guard condition (cond attribute) evaluates to true.

<count>

An optional or repeated grammar rule.

Parent Tags

<count> <item> <rule>

Content Syntax

The value of this tag may be empty, or may contain one or more elements, where an element is either parsed text or one of the following tags.

<count> <item> <one-of> <ruleref> <token>

Attributes

Name	Req/opt	Default	Valid values
number	Optional		*Text*
tag	Optional		*Text*
xml:lang	Optional		*Text*

Compatibility

VoiceXML 1.0 No, 2.0 Yes.

Description

Specifies that a grammar rule should repeat:

a) Zero or one times (i.e, is optional), which is specified by setting number="?" or number="optional", or

b) Zero or more times, specified by setting number="0+", or

c) One or more times, specified by setting number="1+".

\<disconnect\>

Hang up and end session.

Parent Tags

\<block\> \<catch\> \<error\> \<filled\> \<help\> \<if\> \<noinput\> \<nomatch\>

Content Syntax

The value of this tag must be empty.

Attributes

This tag has no attributes.

Compatibility

VoiceXML 1.0 Yes, 2.0 Yes.

Description

Disconnects the call and terminates the browser session.

<div>

Obsolete. *Mark sentence or paragraph.*

Parent Tags

<audio> <choice> <div> <emp> <enumerate> <prompt> <pros>

Content Syntax

The value of this tag may be empty, or may contain one or more elements, where an element is either parsed text or one of the following tags.

<audio> <break> <div> <emp> <enumerate> <pros> <sayas> <value>

Attributes

Name	Req/opt	Default	Valid values
type	Optional		Text

Compatibility

VoiceXML 1.0 Yes, 2.0 No.

Description

Identifies the text as a sentence or paragraph. Not supported in VoiceXML 2.0.

<dtmf>

Obsolete. *Touch-tone grammar.*

Parent Tags

<field> <form> <link> <transfer>

Content Syntax

The value of this tag must be parsed text.

Attributes

Name	Req/opt	Default	Valid values
caching	Optional	"fast"	"fast" "safe"
fetchhint	Optional	"safe"	"prefetch" "safe" "stream"
fetchtimeout	Optional		Text
scope	Optional		"dialog" "document"
src	Optional		Text
type	Optional		Text

Compatibility

VoiceXML 1.0 Yes, 2.0 No.

Description

Specifies a touch-tone grammar. Not supported in VoiceXML 2.0.

<else>
Alternative code.

Parent Tags

<if>

Content Syntax

The value of this tag must be empty.

Attributes

This tag has no attributes.

Compatibility

VoiceXML 1.0 Yes, 2.0 Yes.

Description

Marks the beginning of content to be executed when the parent <if> tag and all <elseif>s at the same level of nesting (if any) have conditions (cond attributes) which evaluate to false. The content ends at the closing </if> tag.

<elseif>

Alternative code.

Parent Tags

<if>

Content Syntax

The value of this tag must be empty.

Attributes

Name	Req/opt	Default	Valid values
cond	Required		Text

Compatibility

VoiceXML 1.0 Yes, 2.0 Yes.

Description

Marks the beginning of content to be executed when a) the parent <if> tag and all preceding <elseif> tags at the same level of nesting (if any) have conditions (cond attributes) which evaluate to false, and b) the cond attribute evaluates to true. The content ends at the next <elseif/> or <else> or closing </if> tag at the same level of nesting, which ever comes first.

\<emp\>
Obsolete. *Emphasize speech.*

Parent Tags
\<audio\> \<choice\> \<div\> \<emp\> \<enumerate\> \<prompt\> \<pros\>

Content Syntax
The value of this tag may be empty, or may contain one or more elements, where an element is either parsed text or one of the following tags.

\<audio\> \<break\> \<div\> \<emp\> \<enumerate\> \<pros\> \<sayas\> \<value\>

Attributes

Name	Req/opt	Default	Valid values
level	Optional	"moderate"	"moderate" "none" "reduced" "strong"

Compatibility
VoiceXML 1.0 Yes, 2.0 No.

Description
Requests that the synthesized speech be emphasized. Not supported by VoiceXML 2.0.

<emphasis>

Emphasize speech.

Parent Tags

<audio> <choice> <emphasis> <enumerate> <p> <paragraph>
<prompt> <prosody> <s> <sentence> <voice>

Content Syntax

The value of this tag may be empty, or may contain one or more
elements, where an element is either parsed text or one of the following
tags.

<audio> <break> <emphasis> <enumerate> <mark> <phoneme>
<prosody> <say-as> <value> <voice>

Attributes

Name	Req/opt	Default	Valid values
level	Optional	"moderate"	"moderate" "none" "reduced" "strong"

Compatibility

VoiceXML 1.0 No, 2.0 Yes.

Description

Requests a specific level of emphasis (weight or stress) of the
synthesized speech which is generated from the text enclosed within the
tag.

<enumerate>

Prompt for options or choices.

Parent Tags

<audio> <block> <catch> <choice> <emphasis> <enumerate> <error>
<field> <filled> <help> <if> <initial> <menu> <noinput> <nomatch>
<object> <p> <paragraph> <prompt> <prosody> <record> <s>
<sentence> <subdialog> <transfer> <voice>

Content Syntax

The value of this tag may be empty, or may contain one or more
elements, where an element is either parsed text or one of the following
tags.

<audio> <break> <emphasis> <enumerate> <mark> <p> <paragraph>
<phoneme> <prosody> <s> <say-as> <sentence> <value> <voice>

Attributes

This tag has no attributes.

Compatibility

VoiceXML 1.0 Yes, 2.0 Yes.

Description

This tag automatically generates a prompt describing menu selections
(see the <choice> tag) or field options (see the <option> tag).

The contents of <enumerate> is a prompt which is executed as if it were
enclosed within a <prompt> tag. This prompt is repeated once for each
choice or option in the order they are written within the menu or field.
Two special variables are available: _prompt, which is replaced by the
value of the <choice> or <option> tag, and _dtmf, which is replaced by
the touch-tone digit which was assigned to that choice or option. To
include _prompt or _dtmf in the prompt, use a <value> tag. If the
<enumerate> tag is empty, the browser constructs a default template for
the prompt. This default is not defined by the VoiceXML language, it is
browser-dependent.

\<error\>

Catch an error event.

Parent Tags

\<field\> \<form\> \<initial\> \<menu\> \<object\> \<record\> \<subdialog\> \<transfer\> \<vxml\>

Content Syntax

The value of this tag may be empty, or may contain one or more elements, where an element is either parsed text or one of the following tags.

\<assign\> \<audio\> \<clear\> \<disconnect\> \<enumerate\> \<exit\> \<goto\> \<if\> \<log\> \<prompt\> \<reprompt\> \<return\> \<script\> \<submit\> \<throw\> \<value\> \<var\>

Attributes

Name	Req/opt	Default	Valid values
cond	Optional		Text
count	Optional		Text

Compatibility

VoiceXML 1.0 Yes, 2.0 Yes.

Description

A short-hand for \<catch event="error"\>, see \<catch\>.

\<example\>
Example of input which matches a grammar rule.

Parent Tags
\<rule\>

Content Syntax
The value of this tag must be parsed text.

Attributes
This tag has no attributes.

Compatibility
VoiceXML 1.0 No, 2.0 Yes.

Description
The value of this tag is an example of an utterance which matches the parent grammar rule.

<exit>

End browser session.

Parent Tags

<block> <catch> <error> <filled> <help> <if> <noinput> <nomatch>

Content Syntax

The value of this tag must be empty.

Attributes

Name	Req/opt	Default	Valid values
expr	Optional		Text
namelist	Optional		Text

Compatibility

VoiceXML 1.0 Yes, 2.0 Yes.

Description

Terminates the browser session. What happens next is platform-dependent, typical is that the browser will disconnect the call. The expr attribute gives an ECMAScript expression which is evaluated and returned to the browser. The namelist attribute gives a whitespace-separated list of variable names, the names and values from which are returned to the browser. What the browser does with these values is browser-dependent.

<field>
Item which collects user input.

Parent Tags
<form>

Content Syntax
The value of this tag may be empty, or may contain one or more elements, where an element is either parsed text or one of the following tags.

<audio> <catch> <enumerate> <error> <filled> <grammar> <help> <link> <noinput> <nomatch> <option> <prompt> <property> <value>

Attributes

Name	Req/opt	Default	Valid values
cond	Optional		Text
expr	Optional		Text
modal	Optional	"false"	"false" "true"
name	Optional		Text
slot	Optional		Text
type	Optional		Text

Compatibility
VoiceXML 1.0 Yes, 2.0 Yes.

Description
Describes an input field in a form. The purpose of a field is to collect a value from the user.

When the field is visited, the browser will first run executable content within the <field> tag (which will typically play a prompt soliciting input from the user), and then wait for user input. This user input is used to set the field variable, as declared by the name attribute.

Acceptable input must be specified either by using the type attribute to specify a built-in type such as "digits", or by specifying one or more <grammar> tags.

The cond and count attributes can be specified to control when the field is eligible to be visited. The prompt counter can be reset to zero and the field variable set to undefined by using the <clear> tag, this makes the field eligible to be visited again.

The translation from the spoken or dialed input to a numerical, string or boolean data value ("semantic interpretation") is done by the browser. For built-in types such as "number", the touch-tone grammars and interpretation are defined by the VoiceXML standard but the speech grammars and interpretation depend on the browser. When a speech grammar is specified instead of a built-in type, the value may default to the string value of the user's utterance as reported by the speech recognizer, or may be set via semantic tags in the grammar. For touch-tone grammars, the value will be the string of touch-tone digits dialed in the absence of semantic tagging. The final VoiceXML 2.0 standard may specify more details of the semantic interpretation than the current draft.

<filled>

Code to run when items are filled in.

Parent Tags

<field> <form> <object> <record> <subdialog> <transfer>

Content Syntax

The value of this tag may be empty, or may contain one or more elements, where an element is either parsed text or one of the following tags.

<assign> <audio> <clear> <disconnect> <enumerate> <exit> <goto> <if> <log> <prompt> <reprompt> <return> <script> <submit> <throw> <value> <var>

Attributes

Name	Req/opt	Default	Valid values
mode	Optional	"ail"	"all" "any"
namelist	Optional		Text

Compatibility

VoiceXML 1.0 Yes, 2.0 Yes.

Description

Contains content to be executed when one or more fields are filled in as a result of user input. A <filled> tag is not triggered when field variables are assigned through ECMAScript statements.

If the <filled> tag is a child of a dialog item such as a <field>, it is triggered when that item has been filled. It is illegal to specify a namelist or mode attribute in this case.

If the <filled> tag is a child of a <form>, then by default it is triggered when all fields in the form have been filled. This can be modified by the namelist and mode attributes. The namelist attribute specifies a list of field names (default, all fields), the mode attribute can be set to "all" or "any" to specify how many of the fields must be filled in.

The typical use for <filled> is to validate user input. If the user input is invalid, the typical action is to play a prompt and use <clear> to re-set the invalid field(s).

\<form>

Dialog for collecting user input.

Parent Tags

\<vxml>

Content Syntax

The value of this tag may be empty or contain one or more of the following tags.

\<block> \<catch> \<error> \<field> \<filled> \<grammar> \<help> \<initial> \<link> \<noinput> \<nomatch> \<object> \<property> \<record> \<subdialog> \<transfer> \<var>

Attributes

Name	Req/opt	Default	Valid values
id	Optional		*Text*
scope	Optional	"dialog"	"dialog" "document"

Compatibility

VoiceXML 1.0 Yes, 2.0 Yes.

Description

A form contains form items, grammars (\<grammar> tags), links (\<link> tags), event handlers, variable declarations (\<var> tags) and properties (\<property> tags). A VoiceXML application is always interpreting exactly one dialog, meaning one \<form> or \<menu>.

Form items include \<block>, \<field>, \<initial>, \<object>, \<property>, \<record>, \<subdialog> and \<transfer>. By default, items are interpreted ("visited") in the order they appear within the form. Visiting an item means executing its content and, in the case of \<field>, \<record> and \<transfer>, waiting for user input. In the case of \<field> and \<record>, user input sets the form item variable. In the case of \<transfer>, grammars can be specified which recognize user input and abort the attempt to transfer the call. An item is visited only if a) its cond attribute evaluates to true and b) its form item variable has the special ECMAScript value called undefined (which is the default initial value of all form item variables).

If there are form-level links (i.e., \<link> tags which are a child of the \<form> tag), then they are active while the form is being interpreted (it is illegal to specify a scope for a grammar inside a link).

A grammar which is a child of the form (i.e., a form-level grammar as opposed to a child of a form item) is active for the entire time that the form is being visited, unless a <field> or <record> tag with the modal attribute set to "true" is being visited, in which case all grammars outside the form item are inactive, including the form-level grammar. A form-level grammar will usually be tagged to indicate that one or more fields are to be set by user input which matches the form. At the time of writing, the tag syntax is not specified by the proposed VoiceXML 2.0 standard and is therefore browser-dependent. An <initial> tag is typically used to provide a prompt for a form-level grammar.

If the scope attribute of the form is set to "document", then form-level grammars are active for the entire time the current page is loaded. If form-level grammars are specified, this is said to be a mixed-initiative or user-directed form. If the form also has document scope, this is said to be a mixed-initiative or user-directed application.

<goto>
Transfer to another dialog and item.

Parent Tags

<block> <catch> <error> <filled> <help> <if> <noinput> <nomatch>

Content Syntax

The value of this tag must be empty.

Attributes

Name	Req/opt	Default	Valid values
caching	Optional		"fast" "safe"
expr	Optional		*Text*
expritem	Optional		*Text*
fetchaudio	Optional		*Text*
fetchhint	Optional		"prefetch" "safe" "stream"
fetchtimeout	Optional		*Text*
next	Optional		*Text*
nextitem	Optional		*Text*

Compatibility

VoiceXML 1.0 Yes, 2.0 Yes.

Description

The <goto> tag is found in executable content. It causes a transition to a new item, which may be in a different dialog and in a different page.

The item is specified by one of the optional nextitem or expritem attributes. It is illegal to specify both. If neither are specified, then the usual criteria in the Form Interpretation Algorithm are used to select the item: the first item in the order written where the cond attribute expression evaluates to true and the item variable is undefined.

The URL of the new dialog is specified by either the next or the expr attribute. If neither is specified, the current dialog is assumed. It is illegal to specify both.

The new dialog is specified by an optional "anchor" (also called "fragment") appended to the URL as "#dialogname". If no dialog name is specified in the URL, then the first dialog on the page is assumed.

If a page is specified in the URL, then the new page is loaded, replacing the current page. The current page is unloaded, losing all variable values, even if the URL specifies the same page that is currently loaded. If the current page is implied by specifying only an anchor in the URL, then the current page variables are preserved.

<grammar>

Speech or touch-tone grammar.

Parent Tags

<choice> <field> <form> <link> <record> <transfer>

Content Syntax

The value of this tag may be empty, or may contain one or more elements, where an element is either parsed text or one of the following tags.

<import> <rule>

Attributes

Name	Req/opt	Default	Valid values
caching	Optional		"fast" "safe"
fetchhint	Optional		"prefetch" "safe" "stream"
fetchtimeout	Optional		Text
mode	Optional	"speech"	"dtmf" "speech"
root	Optional		Text
scope	Optional		"dialog" "document"
src	Optional		Text
type	Optional		Text
version	Optional	"1.0"	Text
xml:lang	Optional		Text

Compatibility

VoiceXML 1.0 Yes, 2.0 Yes.

Description

The <grammar> tag specifies a grammar, which is a template specifying legal user input. A grammar is conceptually similar to a regular expression: it defines a set of strings (sequences of words or tokens such as digits) to match to spoken or dialed input in the same way that a regular expression defines a set of strings to match against text.

Grammars may be specified in-line as the value of the <grammar> tag or externally by using the src attribute to specify a URL.

A compliant VoiceXML 2.0 browser must support the XML form of SRGF.

Browsers may support other grammar formats. In the draft VoiceXML 2.0 standard current at the time of writing, it is not permitted to use the XML form of the SRGF for an in-line grammar, however this may change in the published standard. The type attribute may be used to specify the MIME type of the grammar. The browser may also determine the grammar type from the filename extension or from the HTTP response header.

The scope attribute may be specified for <grammar> tags which are the child of a form, it may be set to "dialog" or "document". Dialog scope means that the grammar is active when the form is being visited, document scope means that the grammar is active for the entire time that the current page is loaded. It is illegal to specify a scope for a grammar which is not the child of a form. Field grammars are active when the field is being visited, link grammars have the scope of the parent of the link. For example, if the link is a child of a form, the grammar is in scope when that form is being visited.

Usually, a grammar which is in scope is also active (that is, user input is monitored for a possible match to that grammar). An exception to this rule occurs when a form item is being visited which has its modal attribute set to "true", in which case a grammar outside that item is not active even it is in scope.

\<help\>

Catch a help event.

Parent Tags

\<field\> \<form\> \<initial\> \<menu\> \<object\> \<record\> \<subdialog\> \<transfer\> \<vxml\>

Content Syntax

The value of this tag may be empty, or may contain one or more elements, where an element is either parsed text or one of the following tags.

\<assign\> \<audio\> \<clear\> \<disconnect\> \<enumerate\> \<exit\> \<goto\> \<if\> \<log\> \<prompt\> \<reprompt\> \<return\> \<script\> \<submit\> \<throw\> \<value\> \<var\>

Attributes

Name	Req/opt	Default	Valid values
cond	Optional		*Text*
count	Optional		*Text*

Compatibility

VoiceXML 1.0 Yes, 2.0 Yes.

Description

A short-hand for \<catch event="help"\>, see \<catch\>.

<if>

Conditional code.

Parent Tags

<block> <catch> <error> <filled> <help> <if> <noinput> <nomatch>

Content Syntax

The value of this tag may be empty, or may contain one or more elements, where an element is either parsed text or one of the following tags.

<assign> <audio> <clear> <disconnect> <else> <elseif> <enumerate> <exit> <goto> <if> <log> <prompt> <reprompt> <return> <script> <submit> <throw> <value> <var>

Attributes

Name	Req/opt	Default	Valid values
cond	Required		Text

Compatibility

VoiceXML 1.0 Yes, 2.0 Yes.

Description

The <if> tag is used to create conditional executable content. The ECMAScript expression specified by the cond tag is evaluated. If the result is true, then the content between the <if> and the first <else>, <elseif> or </if> ("endif") at the same level of nesting is executed. If the result is false, execution skips to the first <else>, <elseif> or </if> at the same level of nesting.

<import>
Alias name for a grammar rule.

Parent Tags

<grammar>

Content Syntax

The value of this tag must be empty.

Attributes

Name	Req/opt	Default	Valid values
name	Required		Text
uri	Required		Text

Compatibility

VoiceXML 1.0 No, 2.0 Yes.

Description

Defines a convenient short-hand name (alias) for an externally defined grammar or grammar rule. The name attribute specifies the alias, the uri attribute specifies the URL of the grammar or grammar rule. A rule within a grammar is specified using the "anchor" or "fragment" notation "#rulename" at the end of the URI. The alias can be used later in the import attribute of the <ruleref> tag. The <import> definition must precede the use of the alias by <ruleref>.

\<initial>

Prompt for form-level input.

Parent Tags

\<form>

Content Syntax

The value of this tag may be empty, or may contain one or more elements, where an element is either parsed text or one of the following tags.

\<audio> \<catch> \<enumerate> \<error> \<help> \<link> \<noinput> \<nomatch> \<prompt> \<property> \<value>

Attributes

Name	Req/opt	Default	Valid values
cond	Optional		Text
expr	Optional		Text
name	Optional		Text

Compatibility

VoiceXML 1.0 Yes, 2.0 Yes.

Description

The \<initial> tag specifies a form item which has prompts, a prompt counter and event handlers but no grammars or \<filled> actions. It is typically used to provide a prompt which solicits input to match form-level grammars.

\<item>
Group part of a grammar.

Parent Tags

\<count> \<item> \<one-of> \<rule>

Content Syntax

The value of this tag may be empty, or may contain one or more elements, where an element is either parsed text or one of the following tags.

\<count> \<item> \<one-of> \<ruleref> \<token>

Attributes

Name	Req/opt	Default	Valid values
tag	Optional		Text
weight	Optional		Text
xml:lang	Optional		Text

Compatibility

VoiceXML 1.0 No, 2.0 Yes.

Description

An \<item> tag is used to group a part of a grammar rule into a single unit. The \<item>...\</item> tags play roles similar to open and closing parentheses (...) in an arithmetic expression.

\<link\>

Hyperlink or throw event when input matches.

Parent Tags

\<field\> \<form\> \<initial\> \<vxml\>

Content Syntax

The value of this tag may be empty, or may contain one or more \<grammar\> tags.

Attributes

Name	Req/opt	Default	Valid values
caching	Optional		"fast" "safe"
event	Optional		Text
expr	Optional		Text
fetchaudio	Optional		Text
fetchhint	Optional		"prefetch" "safe" "stream"
fetchtimeout	Optional		Text
next	Optional		Text

Compatibility

VoiceXML 1.0 Yes, 2.0 Yes.

Description

The \<link\> tag specifies a link. It must contain one or more grammars. When a grammar matches, then either a hyperlink is executed or an event is thrown.

Either a URL for the hyperlink is specified by the uri or expr attributes, or an event is specified by using the event attribute. Exactly one of the event, uri or expr attributes must be specified.

A hyperlink is executed as for a \<goto\> tag (except that you cannot specify a specific dialog item as the target of the link because the \<link\> tag has no nexitem attribute).

If an event is thrown, the event is considered to be thrown from the dialog item which is currently being visited, so an event handler is sought starting from that item and the scope of variables is defined by the scope of that item.

Grammars in a <link> tag are always scoped to the link, it is illegal to specify the scope attribute of a <grammar> tag in a link. The scope of the link (and therefore the scope of any grammars in the link) is the parent tag, so for example if the <link> tag is a child of a form, then the link has dialog scope.

\<log\>

Output a debug message.

Parent Tags

\<block\> \<catch\> \<error\> \<filled\> \<help\> \<if\> \<noinput\> \<nomatch\>

Content Syntax

The value of this tag may be empty, or may contain one or more elements, where an element is either parsed text or a \<value\> tag.

Attributes

This tag has no attributes.

Compatibility

VoiceXML 1.0 No, 2.0 Yes.

Description

The \<log\> tag writes its content to a debug trace file. What exactly this means (e.g., the location of the file and how to read it) is browser-dependent.

<mark>

Notification when TTS reaches this point.

Parent Tags

<audio> <choice> <emphasis> <enumerate> <p> <paragraph>
<prompt> <prosody> <s> <sentence> <voice>

Content Syntax

The value of this tag must be empty.

Attributes

Name	Req/opt	Default	Valid values
name	Required		Text

Compatibility

VoiceXML 1.0 No, 2.0 Yes.

Description

The <mark> tag is used to notify an application in real-time that the
speech synthesizer has reached a particular point in the text. It will
usually be ignored by VoiceXML browsers.

<menu>

Dialog for making a selection between fixed choices.

Parent Tags

<vxml>

Content Syntax

The value of this tag may be empty, or may contain one or more elements, where an element is either parsed text or one of the following tags.

<audio> <catch> <choice> <enumerate> <error> <help> <noinput> <nomatch> <prompt> <property> <value>

Attributes

Name	Req/opt	Default	Valid values
accept	Optional	"exact"	"approximate" "exact"
dtmf	Optional	"false"	"false" "true"
id	Optional		Text
scope	Optional	"dialog"	"dialog" "document"

Compatibility

VoiceXML 1.0 Yes, 2.0 Yes.

Description

A menu is a dialog which presents a choice to the user. It is a syntactic short-hand for a form with one field. Within the menu, the <choice> tag plays the role of an <option> tag in that field.

The <menu> tag can contain a prompt. Typically this prompt will ask the user to make a selection from the supplied choices.

Each <choice> tag specifies a grammar and/or touch-tone digit. When user input matches a choice, an event is thrown or a transition is made to a new dialog (and perhaps a new page) based on the event, uri or expr attributes of the <choice> tag.

The <enumerate> tag can be used to generate all or part of the prompt automatically based on the choices.

If the dtmf attribute of <menu> is set to "true", touch-tone digits will be assigned to <choice> tags which do not set their own dtmf attribute.

\<meta\>
Page properties.

Parent Tags
\<vxml\>

Content Syntax
The value of this tag must be empty.

Attributes

Name	Req/opt	Default	Valid values
content	Required		Text
http-equiv	Optional		Text
name	Optional		Text

Compatibility
VoiceXML 1.0 Yes, 2.0 Yes.

Description
The \<meta\> tag is similar to the tag of the same name in HTML, it provides information about the page as a name/value pair. The value is specified in the content attribute. The name either corresponds to an HTTP response header field such as "Expires" or "Date", in which case the http-equiv attribute is specified, or a meta-data document property, in which case name is used.

HTTP header fields specified by http-equiv override fields which are sent by the server, which can be used for example to change the expiration date for the cache.

The VoiceXML standard recommends, but does not mandate, that properties with the following names be supported.

name="author", name of the author.

name="maintainer", the author's e-mail address. This might be used to generate a pager or e-mail notification in the event of an error on this page.

name="copyright", an intellectual property legal notice.

name="keywords", a set of keywords for the page.

name="description", a short description of the page for human readers.

name="robots", a description of the page formatted for search engines.

Browsers ignore unrecognized meta tags.

\<noinput\>

Catch a noinput event.

Parent Tags

\<field\> \<form\> \<initial\> \<menu\> \<object\> \<record\> \<subdialog\>
\<transfer\> \<vxml\>

Content Syntax

The value of this tag may be empty, or may contain one or more
elements, where an element is either parsed text or one of the following
tags.

\<assign\> \<audio\> \<clear\> \<disconnect\> \<enumerate\> \<exit\> \<goto\> \<if\>
\<log\> \<prompt\> \<reprompt\> \<return\> \<script\> \<submit\> \<throw\>
\<value\> \<var\>

Attributes

Name	Req/opt	Default	Valid values
cond	Optional		Text
count	Optional		Text

Compatibility

VoiceXML 1.0 Yes, 2.0 Yes.

Description

A short-hand for \<catch event="noinput"\>, see \<catch\>.

<nomatch>
Catch a nomatch event.

Parent Tags

<field> <form> <initial> <menu> <object> <record> <subdialog> <transfer> <vxml>

Content Syntax

The value of this tag may be empty, or may contain one or more elements, where an element is either parsed text or one of the following tags.

<assign> <audio> <clear> <disconnect> <enumerate> <exit> <goto> <if> <log> <prompt> <reprompt> <return> <script> <submit> <throw> <value> <var>

Attributes

Name	Req/opt	Default	Valid values
cond	Optional		Text
count	Optional		Text

Compatibility

VoiceXML 1.0 Yes, 2.0 Yes.

Description

A short-hand for <catch event="nomatch">, see <catch>.

<object>
Execute a browser extension.

Parent Tags

<form>

Content Syntax

The value of this tag may be empty, or may contain one or more elements, where an element is either parsed text or one of the following tags.

<audio> <catch> <enumerate> <error> <filled> <help> <noinput> <nomatch> <param> <prompt> <property> <value>

Attributes

Name	Req/opt	Default	Valid values
archive	Optional		*Text*
caching	Optional		"fast" "safe"
classid	Optional		*Text*
codebase	Optional		*Text*
codetype	Optional		*Text*
cond	Optional		*Text*
data	Optional		*Text*
expr	Optional		*Text*
fetchhint	Optional		"prefetch" "safe" "stream"
fetchtimeout	Optional		*Text*
name	Optional		*Text*
type	Optional		*Text*

Compatibility

VoiceXML 1.0 Yes, 2.0 Yes.

Description

The <object> tag is a form item which invokes a browser-specific extension. If a browser visits an object tag that it does not support, it must throw an error.unsupported.object event.

If the object is successfully invoked, it sets the item variable (as declared by the name attribute), multiple values may be returned by setting multiple properties of that variable.

<one-of>
Set of alternatives in a grammar.

Parent Tags

<count> <item> <rule>

Content Syntax

The value of this tag may be empty, or may contain one or more <item> tags.

Attributes

Name	Req/opt	Default	Valid values
tag	Optional		*Text*
xml:lang	Optional		*Text*

Compatibility

VoiceXML 1.0 No, 2.0 Yes.

Description

Defines a choice in a grammar. Exactly one of the enclosed <item> tags must match.

\<option\>
Defines a field selection.

Parent Tags
\<field\>

Content Syntax
The value of this tag must be parsed text.

Attributes

Name	Req/opt	Default	Valid values
dtmf	Optional		Text
value	Optional		Text

Compatibility
VoiceXML 1.0 Yes, 2.0 Yes.

Description
Specifying \<option\> tags is an alternative to providing a grammar when the input to a field must be one of a fixed list of alternatives. Each \<option\> tag specifies an alternative.

You can specify both grammar(s) and option(s) in a field, in which case all are active.

The value attribute specifies the value to assign to the field variable when the user selects this choice. If not specified, the value of the tag (i.e., the text inside the tag) is used, with leading and trailing white space removed.

The dtmf attribute specifies a touch-tone digit which selects this option.

The text values of the \<option\> tags within a field are combined automatically to create a grammar which matches the choices.

The \<enumerate\> tag can be used to generate a prompt which lists the options in a field.

\<p>

Synonym for \<paragraph>.

Parent Tags

\<audio> \<choice> \<enumerate> \<prompt> \<prosody> \<voice>

Content Syntax

The value of this tag may be empty, or may contain one or more elements, where an element is either parsed text or one of the following tags.

\<audio> \<break> \<emphasis> \<enumerate> \<mark> \<phoneme> \<prosody> \<s> \<say-as> \<sentence> \<value> \<voice>

Attributes

Name	Req/opt	Default	Valid values
xml:lang	Optional		Text

Compatibility

VoiceXML 1.0 No, 2.0 Yes.

Description

Synonym for \<paragraph>. See \<paragraph>.

\<paragraph>

Paragraph marker for TTS.

Parent Tags

\<audio> \<choice> \<enumerate> \<prompt> \<prosody> \<voice>

Content Syntax

The value of this tag may be empty, or may contain one or more elements, where an element is either parsed text or one of the following tags.

\<audio> \<break> \<emphasis> \<enumerate> \<mark> \<phoneme> \<prosody> \<s> \<say-as> \<sentence> \<value> \<voice>

Attributes

Name	Req/opt	Default	Valid values
xml:lang	Optional		Text

Compatibility

VoiceXML 1.0 No, 2.0 Yes.

Description

The \<paragraph> tag is used to enclose a paragraph of text. This may assist the speech synthesizer in determining the cadence (prosody) and interpretation (normalization) of the text.

\<param\>
Parameter of an object or subdialog.

Parent Tags

\<object\> \<subdialog\>

Content Syntax

The value of this tag must be empty.

Attributes

Name	Req/opt	Default	Valid values
expr	Optional		Text
name	Required		Text
type	Optional		Text
value	Optional		Text
valuetype	Optional	"data"	"data" "ref"

Compatibility

VoiceXML 1.0 Yes, 2.0 Yes.

Description

The \<param\> tag defines a parameter to be passed to an object or subdialog. It is similar to a function or subroutine argument. In the case of a subdialog, the name of the parameter (specified by the name attribute) must correspond to the name of a variable defined in the dialog which is being called, this variable is set to the value of the parameter (specified by the value or expr attributes).

If the parameter is being passed to an object, the valuetype attribute may be set to "ref" or "data" to indicate if the value in the data attribute is an URL which points to the value ("ref"), or is itself the value to use ("data").

\<phoneme\>
Phonetic pronunciation.

Parent Tags

\<audio\> \<choice\> \<emphasis\> \<enumerate\> \<p\> \<paragraph\>
\<prompt\> \<prosody\> \<s\> \<sentence\> \<voice\>

Content Syntax

The value of this tag must be parsed text.

Attributes

Name	Req/opt	Default	Valid values
alphabet	Optional		Text
ph	Required		Text

Compatibility

VoiceXML 1.0 No, 2.0 Yes.

Description

Specifies exact pronunciation using a phonetic alphabet. The alphabet tag specifies which alphabet to use, which can be "ipa", "worldbet", "xsampa" or a browser-specific value. The ph attribute specifies the phonemes using the given alphabet.

The value of the tag (text enclosed within the tag) provides alternative text which can be used by a graphical browser, for example. A VoiceXML browser will ignore the value or might use the value in the event that the requested alphabet is unsupported (the proposed VoiceXML 2.0 standard does not specify how alternative text should be used).

\<prompt\>

Speak text using TTS.

Parent Tags

\<block\> \<catch\> \<error\> \<field\> \<filled\> \<help\> \<if\> \<initial\> \<menu\>
\<noinput\> \<nomatch\> \<object\> \<record\> \<subdialog\> \<transfer\>

Content Syntax

The value of this tag may be empty, or may contain one or more elements, where an element is either parsed text or one of the following tags.

\<audio\> \<break\> \<emphasis\> \<enumerate\> \<mark\> \<p\> \<paragraph\>
\<phoneme\> \<prosody\> \<s\> \<say-as\> \<sentence\> \<value\> \<voice\>

Attributes

Name	Req/opt	Default	Valid values
bargein	Optional		"false" "true"
cond	Optional		Text
count	Optional		Text
timeout	Optional		Text
xml:lang	Optional		Text

Compatibility

VoiceXML 1.0 Yes, 2.0 Yes.

Description

The \<prompt\> tag speaks its value (the text enclosed within the tag) using speech synthesis.

The cond attribute may be used to give an expression which must evaluate to true for the prompt to be played.

The count attribute may be used to select which prompt within a item will be played based on the prompt counter (the number of times the user has been prompted in that item).

The timeout attribute specifies how long to wait for input after this prompt has been played.

The xml:lang attribute specifies the language of the enclosed text. If the browser does not support this language, it will throw an error event.

The bargein attribute specifies whether user input will interrupt the prompt and be recognized. A VoiceXML browser is not obligated to support bargein.

<property>
Platform setting.

Parent Tags

<field> <form> <initial> <menu> <object> <record> <subdialog>
<transfer> <vxml>

Content Syntax

The value of this tag must be empty.

Attributes

Name	Req/opt	Default	Valid values
name	Required		Text
value	Required		Text

Compatibility

VoiceXML 1.0 Yes, 2.0 Yes.

Description

The <property> tag sets a named value called a property within a page,
dialog or dialog item. Properties are typically used to override built-in
language default attributes or to set browser-specific options.

<pros>
Obsolete. *Set prosody.*

Parent Tags

<audio> <choice> <div> <emp> <enumerate> <prompt> <pros>

Content Syntax

The value of this tag may be empty, or may contain one or more elements, where an element is either parsed text or one of the following tags.

<audio> <break> <div> <emp> <enumerate> <pros> <sayas> <value>

Attributes

Name	Req/opt	Default	Valid values
pitch	Optional		Text
range	Optional		Text
rate	Optional		Text
vol	Optional		Text

Compatibility

VoiceXML 1.0 Yes, 2.0 No.

Description

Specifies the prosody of spoken text. Replaced by <prosody> in VoiceXML 2.0.

\<prosody>
Set prosody.

Parent Tags
\<audio> \<choice> \<emphasis> \<enumerate> \<p> \<paragraph>
\<prompt> \<prosody> \<s> \<sentence> \<voice>

Content Syntax
The value of this tag may be empty, or may contain one or more elements, where an element is either parsed text or one of the following tags.

\<audio> \<break> \<emphasis> \<enumerate> \<mark> \<p> \<paragraph>
\<phoneme> \<prosody> \<s> \<say-as> \<sentence> \<value> \<voice>

Attributes

Name	Req/opt	Default	Valid values
contour	Optional		Text
duration	Optional		Text
pitch	Optional		Text
range	Optional		Text
rate	Optional		Text
volume	Optional		Text

Compatibility
VoiceXML 1.0 No, 2.0 Yes.

Description
The \<prosody> tag specifies the prosody of the text enclosed within the tag. Prosody refers to the pitch (average frequency), pitch range (difference between the maximum and minimum pitch), rate (words per minute) and volume (loudness) of the speech. If an attribute is not specified, the value remains as it was previously set, it is not reset to a default. The new values apply to all the text enclosed within the tag (unless overridden by a nested \<prosody> tag) and then reverts to the previous values when the end tag is reached.

VoiceXML browsers are not obligated to fully implement all speech markup tags, however, a compliant browser must silently ignore any tags it does not support, and should make its best efforts to map any unimplemented mark-up to something similar that is supported.

<record>
Record audio.

Parent Tags

<form>

Content Syntax

The value of this tag may be empty, or may contain one or more elements, where an element is either parsed text or one of the following tags.

<audio> <catch> <enumerate> <error> <filled> <grammar> <help> <noinput> <nomatch> <prompt> <property> <value>

Attributes

Name	Req/opt	Default	Valid values
beep	Optional	"false"	"false" "true"
cond	Optional		Text
dtmfterm	Optional	"true"	"false" "true"
expr	Optional		Text
finalsilence	Optional		Text
maxtime	Optional		Text
modal	Optional	"true"	"false" "true"
name	Optional		Text
type	Optional		Text

Compatibility

VoiceXML 1.0 Yes, 2.0 Yes.

Description

The <record> tag is a form item which records user input to its item variable (as declared by the name attribute). This audio can be played back using <audio> and copied to a server URL by using <submit>.

The type attribute specifies the MIME type of the recorded audio file, a typical example would be type="audio/wav" for a Windows Wave file. The set of formats supported is browser-dependent,

If the browser supports simultaneous recording and speech recognition, then a speech grammar may be specified which will terminate the recording. All browsers should support use of a touch-tone grammar,

which may be used for controlling exactly which digits will be accepted
(as opposed to setting the dtmfterm attribute to "true", which will accept
all digits and may therefore be more vulnerable to talk-off).

If a form- or document-level grammar is matched during the recording,
the recording is terminated, the audio is discarded and execution
transitions according to the usual processing of the event or hyperlink.
Grammars outside of the <record> tag can be disabled by setting the
modal attribute to "true".

\<reprompt\>

Tell browser to prompt for item before collecting input.

Parent Tags

\<block\> \<catch\> \<error\> \<filled\> \<help\> \<if\> \<noinput\> \<nomatch\>

Content Syntax

The value of this tag must be empty.

Attributes

This tag has no attributes.

Compatibility

VoiceXML 1.0 Yes, 2.0 Yes.

Description

The \<reprompt\> tag is used inside executable content to set a flag in the browser to indicate that a new attempt should be made to issue a prompt for the current item. This is intended primarily for use in event handlers. When an event is thrown, the browser assumes that the handler provides a prompt and therefore skips straight to the collect phase when the event handler finishes. Setting this flag overrides that assumption and causes the browser to issue a new prompt before collecting input.

<return>
Return from a subdialog.

Parent Tags
<block> <catch> <error> <filled> <help> <if> <noinput> <nomatch>

Content Syntax
The value of this tag must be empty.

Attributes

Name	Req/opt	Default	Valid values
event	Optional		Text
namelist	Optional		Text

Compatibility
VoiceXML 1.0 Yes, 2.0 Yes.

Description
The <return> tag is used to return from a dialog which was entered as a result of visiting a <subdialog> item.

If the event attribute is specified, then the given event will be thrown immediately after returning.

The namelist attribute specifies a list of whitespace-separated variable names. Each variable name should correspond to a field in the subdialog, upon return the item variable in the <subdialog> will be assigned one property for each name. So for example given <subdialog name="x" src="#aform">, and <return namelist="field1 field2"/>, then following the return x will have two properties x.field1 and x.field2 which are set to the values collected in the subdialog fields.

\<rule\>

Define grammar rule.

Parent Tags

\<grammar\>

Content Syntax

The value of this tag may be empty, or may contain one or more elements, where an element is either parsed text or one of the following tags.

\<count\> \<example\> \<item\> \<one-of\> \<ruleref\> \<token\>

Attributes

Name	Req/opt	Default	Valid values
id	Required		*Text*
scope	Optional	"private"	"private" "public"

Compatibility

VoiceXML 1.0 No, 2.0 Yes.

Description

The \<rule\> tag encloses a named rule in a grammar. The rule is assigned the name given in its id attribute.

A \<ruleref\> tag can be used to reference the rule name in the \<ruleref\>'s src attribute, this causes the rule to be inserted at that point in the grammar.

The \<rule\> tag defines a rule but does not insert it (like defining a subroutine), the \<ruleref\> tag inserts the rule (like calling a subroutine).

<ruleref>

Insert grammar rule.

Parent Tags

<count> <item> <rule>

Content Syntax

The value of this tag must be empty.

Attributes

Name	Req/opt	Default	Valid values
import	Optional		*Text*
special	Optional		*Text*
tag	Optional		*Text*
uri	Optional		*Text*
xml:lang	Optional		*Text*

Compatibility

VoiceXML 1.0 No, 2.0 Yes.

Description

The <ruleref> tag inserts a rule into a grammar. The rule is defined elsewhere using the <rule> tag. The rule can be defined by giving a URL in the id attribute, an alias in the import attribute or a special rule in the special attribute.

The uri attribute of <ruleref> must match the id attribute of the <rule>.

The rule may be found in the same grammar by using the fragment notation "#rulename" for the URL.

The rule may be found in an external grammar, i.e. a grammar in a different file, by using the notation "grammarURL#rulename".

If no rulename is specified, the root rule for the grammar is assumed.

Note that SRGF does not permit more than one grammar to be specified per file, the fragment notation refers to a rule within a grammar, not a grammar within a page.

For an explanation of import aliases, see the <import> tag.

Special rules which can be named in the special attribute include "#VOID", which never matches, "#NULL", which always matches, and "#GARBAGE", which matches any part of an utterance.

`<s>`

Synonym for `<sentence>`.

Parent Tags

`<audio>` `<choice>` `<enumerate>` `<p>` `<paragraph>` `<prompt>` `<prosody>` `<voice>`

Content Syntax

The value of this tag may be empty, or may contain one or more elements, where an element is either parsed text or one of the following tags.

`<audio>` `<break>` `<emphasis>` `<enumerate>` `<mark>` `<phoneme>` `<prosody>` `<say-as>` `<value>` `<voice>`

Attributes

Name	Req/opt	Default	Valid values
xml:lang	Optional		Text

Compatibility

VoiceXML 1.0 No, 2.0 Yes.

Description

This is a synonym for the `<sentence>` tag. See `<sentence>`.

\<say-as\>
Speak value.

Parent Tags

\<audio\> \<choice\> \<emphasis\> \<enumerate\> \<p\> \<paragraph\>
\<prompt\> \<prosody\> \<s\> \<sentence\> \<voice\>

Content Syntax

The value of this tag must be parsed text.

Attributes

Name	Req/opt	Default	Valid values
sub	Optional		*Text*
type	Required		"acronym" "address" "currency" "date" "date:d" "date:dmy" "date:m" "date:md" "date:mdy" "date:my" "date:y" "date:ym" "date:ymd" "duration" "duration:h" "duration:hm" "duration:hms" "duration:m" "duration:ms" "duration:s" "measure" "name" "net" "number" "number:digits" "number:ordinal" "telephone" time" "time:h" "time:hm" "time:hms"

Compatibility

VoiceXML 1.0 No, 2.0 Yes.

Description

The \<say-as\> tag specifies how to interpret text for speech synthesis.

The text is either specified as the value of the tag (i.e., the text between the start and end tag), or by the sub attribute. If the sub attribute is specified, the value of the tag is ignored.

The type attribute specifies the data type of the text to be spoken. Note that type="acronym", perhaps counter-intuitively, means to speak a group of letters as a word, as in DEC or NASA, rather than the default which is to spell out the letters, as in IBM or USA.

\<sayas\>
Obsolete. *Speak value.*

Parent Tags

\<audio\> \<choice\> \<div\> \<emp\> \<enumerate\> \<prompt\> \<pros\>

Content Syntax

The value of this tag must be parsed text.

Attributes

Name	Req/opt	Default	Valid values
class	Optional		Text
phon	Optional		Text
sub	Optional		Text

Compatibility

VoiceXML 1.0 Yes, 2.0 No.

Description

This tag was replaced by \<say-as\> in VoiceXML 2.0.

\<script\>
Run ECMAScript code.

Parent Tags
\<block\> \<catch\> \<error\> \<filled\> \<help\> \<if\> \<noinput\> \<nomatch\>
\<vxml\>

Content Syntax
The value of this tag must be parsed text.

Attributes

Name	Req/opt	Default	Valid values
caching	Optional		"fast" "safe"
charset	Optional		Text
fetchhint	Optional		"prefetch" "safe" "stream"
fetchtimeout	Optional		Text
src	Optional		Text

Compatibility
VoiceXML 1.0 Yes, 2.0 Yes.

Description
The \<script\> tag is found in executable content, it specifies ECMAScript code to be run. The code can either be in-line, in other words given as the value of the tag (the text between the start and end tag), or externally via the src attribute which gives a URL.

Note that in-line scripts are parsed by the XML processor looking for the reserved characters "<" and "&", which can appear in ECMAScript expressions or comments. These can be escaped as "<" and "&". Alternatively you can use a CDATA section (provided you are certain that the three-character sequence "]]>" does not occur in the body of your script).

\<sentence\>

Mark a sentence.

Parent Tags

\<audio\> \<choice\> \<enumerate\> \<p\> \<paragraph\> \<prompt\> \<prosody\> \<voice\>

Content Syntax

The value of this tag may be empty, or may contain one or more elements, where an element is either parsed text or one of the following tags.

\<audio\> \<break\> \<emphasis\> \<enumerate\> \<mark\> \<phoneme\> \<prosody\> \<say-as\> \<value\> \<voice\>

Attributes

Name	Req/opt	Default	Valid values
xml:lang	Optional		Text

Compatibility

VoiceXML 1.0 No, 2.0 Yes.

Description

The \<sentence\> tag is used to enclose a sentence of text. This may assist the speech synthesizer in determining the cadence (prosody) and interpretation (normalization) of the text.

<subdialog>

Invoke second dialog.

Parent Tags

<form>

Content Syntax

The value of this tag may be empty, or may contain one or more elements, where an element is either parsed text or one of the following tags.

<audio> <catch> <enumerate> <error> <filled> <help> <noinput> <nomatch> <param> <prompt> <property> <value>

Attributes

Name	Req/opt	Default	Valid values
caching	Optional		"fast" "safe"
cond	Optional		Text
enctype	Optional	"application/x-www-form-urlencoded"	Text
expr	Optional		Text
fetchaudio	Optional		Text
fetchhint	Optional		"prefetch" "safe" "stream"
fetchtimeout	Optional		Text
method	Optional	"get"	"get" "post"
name	Optional		Text
namelist	Optional		Text
src	Required		Text

Compatibility

VoiceXML 1.0 Yes, 2.0 Yes.

Description

The <subdialog> tag is a form item which invokes a call to a second dialog. The called dialog can collect fields and return values.

The URL of the second dialog is specified by the src attribute. The namelist and method attributes can be used to specify variables to be sent and how to send them (similar to the <submit> tag), this may be used only when the URL specifies a new page.

Parameters can be passed to the second dialog by using the <param> tag, each parameter must correspond to a variable declared within the second dialog by using a <var> tag.

The second dialog returns control to the <subdialog> item by using the <return> tag. The <return> tag can specify a namelist attribute which specifies one or more fields in the second dialog. Each field in the namelist is used to set a property of the calling <subdialog> item's variable. The <subdialog> item variable is declared by using the name property.

<submit>

Submit form values, get new page.

Parent Tags

<block> <catch> <error> <filled> <help> <if> <noinput> <nomatch>

Content Syntax

The value of this tag must be empty.

Attributes

Name	Req/opt	Default	Valid values
caching	Optional		"fast" "safe"
enctype	Optional	"application/x-www-form-urlencoded"	Text
expr	Optional		Text
fetchaudio	Optional		Text
fetchhint	Optional		"prefetch" "safe" "stream"
fetchtimeout	Optional		Text
method	Optional	"get"	"get" "post"
namelist	Optional		Text
next	Optional		Text

Compatibility

VoiceXML 1.0 Yes, 2.0 Yes.

Description

The <submit> tag is found in executable content. It fetches a new page from the server and transfers control to a dialog on the new page.

The URL of the new page is specified by the next attribute or the expr attribute, exactly one of these two attributes must be provided.

The namelist attribute specifies which form variables are to be sent. By default, all named items in the current form are sent. If an item variable has properties, the properties will be sent also using the syntax itemname.propertyname=value.

The method attribute is set to "get" or "post" to specify the HTTP command to use in fetching the new page. We recommend using "post" where possible since the HTTP GET command typically limits the amount of form variable data which can be sent in a URL, the limit depends on the server.

\<throw\>

Trigger an event.

Parent Tags

\<block\> \<catch\> \<error\> \<filled\> \<help\> \<if\> \<noinput\> \<nomatch\>

Content Syntax

The value of this tag must be empty.

Attributes

Name	Req/opt	Default	Valid values
event	Required		Text

Compatibility

VoiceXML 1.0 Yes, 2.0 Yes.

Description

The \<throw\> tag throws the event specified by the event attribute. See also \<catch\>.

<token>

Input element in grammar.

Parent Tags

<count> <item> <rule>

Content Syntax

The value of this tag must be parsed text.

Attributes

Name	Req/opt	Default	Valid values
lexicon	Optional		Text
xml:lang	Optional		Text

Compatibility

VoiceXML 1.0 No, 2.0 Yes.

Description

The <token> tag is used to enclose a single recognized unit of speech in a speech grammar. Typically a token is a single word and does not need to be enclosed in a <token> tag. If the token has whitespace it must be enclosed in a <token> tag, which plays the role of quotes "..." in the ABNF form of the grammar language. Using a <token> tag can also be used to specify an alternative language or lexicon for the token by using the xml:lang and lexicon attributes.

<transfer>

Transfer call.

Parent Tags

<form>

Content Syntax

The value of this tag may be empty, or may contain one or more elements, where an element is either parsed text or one of the following tags.

<audio> <catch> <enumerate> <error> <filled> <grammar> <help>
<noinput> <nomatch> <prompt> <property> <value>

Attributes

Name	Req/opt	Default	Valid values
bridge	Optional	"false"	"false" "true"
cond	Optional		Text
connecttimeout	Optional		Text
dest	Optional		Text
destexpr	Optional		Text
expr	Optional		Text
maxtime	Optional		Text
name	Optional		Text

Compatibility

VoiceXML 1.0 Yes, 2.0 Yes.

Description

The <transfer> tag is a form item which instructs the browser to attempt to transfer the call to the URL specified in the dest or destexpr attributes, exactly one of which must be specified. The "phone://" scheme is used to specify PSTN phone numbers, for example dest="phone://8005551212".

By default, the browser completes the transfer as soon as the new call is answered. This leaves the user in a conversation with the called party. The browser, which is now disconnected from the user, throws a telephone.disconnect.transfer event.

If the bridge attribute is set to "true", the browser stays connected after the call is completed.

During the call progress analysis phase, user input is monitored and if it matches a grammar the transfer will be aborted. The <transfer> tag is modal, grammars outside the tag are always inactive while the tag is being visited even if they are in scope.

When the transfer is completed, the item variable (declared by the name attribute) is set to one of the following values indicating the result of the call progress analysis: "busy", "noanswer", "network_busy", "near_end_disconnect" (user hung up or user input matched a grammar), "far_end_disconnect" (called party hung up), or "network_disconnect" (network hung up). Any <filled> tags under the <transfer> tag are executed when the transfer is completed.

<value>

Insert value of expression as text.

Parent Tags

<audio> <block> <catch> <choice> <emphasis> <enumerate> <error>
<field> <filled> <help> <if> <initial> <log> <menu> <noinput> <nomatch>
<object> <p> <paragraph> <prompt> <prosody> <record> <s>
<sentence> <subdialog> <transfer> <voice>

Content Syntax

The value of this tag must be empty.

Attributes

Name	Req/opt	Default	Valid values
audiobase	Optional		Text
class	Optional		Text
expr	Required		Text
mode	Optional	"tts"	"recorded" "tts"

Compatibility

VoiceXML 1.0 Yes, 2.0 Yes.

Description

The <value> tag inserts a value into text, usually text which is the source
of a speech synthesis prompt. It can also be used in text to output by the
<log> tag.

The expression in the expr tag is evaluated to determine the value to
insert.

By default, the literal value of the text is inserted, so <prompt>the quick
<value expr="'brown'"/> fox</prompt> is equivalent to <prompt>the quick
brown fox</prompt>.

If the class attribute is set, this determines how the inserted value is to
be interpreted by the speech synthesizer. For example, if
class="number:ordinal", the value "3" would be spoken as "third" rather
than "three". The values for the class attribute correspond exactly to the
values for the type attribute of the <say-as> tag.

By default, values are spoken using speech synthesis. Some browsers
may support the ability to speak some classes of values by

concatentating pre-recorded vocabulary files (which preferably should be recorded by the same person who recorded other prompts for your application). This is requested by setting the mode attribute to "recorded", though browsers are free to ignore that request.

`<var>`
Declare a variable.

Parent Tags
`<block>` `<catch>` `<error>` `<filled>` `<form>` `<help>` `<if>` `<noinput>`
`<nomatch>` `<vxml>`

Content Syntax
The value of this tag must be empty.

Attributes

Name	Req/opt	Default	Valid values
expr	Optional		Text
name	Required		Text

Compatibility
VoiceXML 1.0 Yes, 2.0 Yes.

Description
The `<var>` tag declares an ECMAScript variable. In VoiceXML, all variables must be declared before use. The name of the variable is given by the name attribute. An initial value can be specified by giving the expr attribute, if not given then the variable is set to the special ECMAScript value called undefined.

The scope of the variable is determined by the parent tag, so for example a `<var>` tag which is the child of a `<form>` has dialog scope.

Variables can be used in expressions in attributes such as expr, expritem, nextexpr etc., and in scripts found in `<script>` tags.

Variables are visible to expressions and scripts in the same scope (e.g., the same dialog) and to outer scopes (in the case of a dialog, that would mean content in the current page, i.e. document scope, and the root document if any, i.e. application scope).

It is legal to declare a variable with the same name as a variable in an outer scope. The variable in the outer scope can still be used by using the syntax scopename.variablename.

It is not legal to reference variables declared in an inner scope, e.g. an expression in document scope cannot reference a variable in dialog scope.

<voice>
TTS voice.

Parent Tags

<audio> <choice> <emphasis> <enumerate> <p> <paragraph>
<prompt> <prosody> <s> <sentence> <voice>

Content Syntax

The value of this tag may be empty, or may contain one or more
elements, where an element is either parsed text or one of the following
tags.

<audio> <break> <emphasis> <enumerate> <mark> <p> <paragraph>
<phoneme> <prosody> <s> <say-as> <sentence> <value> <voice>

Attributes

Name	Req/opt	Default	Valid values
age	Optional		Text
category	Optional		"adult" "child" "elder" "teenager"
gender	Optional		"female" "male" "neutral"
name	Optional		Text
variant	Optional		Text

Compatibility
VoiceXML 1.0 No, 2.0 Yes.

Description
The <voice> tag specifies attributes of the synthesized voice.

If an attribute is not specified, the value remains as it was previously set,
it is not reset to a default. The new values apply to all the text enclosed
within the tag (unless overridden by a nested <voice> tag) and then
reverts to the previous values when the end tag is reached.

VoiceXML browsers are not obligated to fully implement all speech
markup tags, however, a compliant browser must silently ignore any tags
it does not support, and should make its best efforts to map any
unimplemented mark-up to something similar that is supported.

<vxml>
Root tag.

Parent Tags

This is the root tag, so there are no parents.

Content Syntax

The value of this tag must contain one or more of the following tags.

<catch> <error> <form> <help> <link> <menu> <meta> <noinput>
<nomatch> <property> <script> <var>

Attributes

Name	Req/opt	Default	Valid values
application	Optional		Text
base	Optional		Text
version	Required		Text
xml:lang	Optional		Text

Compatibility

VoiceXML 1.0 Yes, 2.0 Yes.

Description

The <vxml> tag is the root tag for a VoiceXML page. All the content of a page is enclosed within one <vxml> tag.

The version tag specifies the VoiceXML version, for VoiceXML 2.0 it must be set to "2.0".

The base attribute specifies the base URL for the page as a hint to the server on how to find resources specified by relative URLs, this works as for HTML.

The application attribute specifies a root document. If the page is loaded as a root document, it is illegal to have an application attribute.

The xml:lang attribute specifies the default language for speech synthesis and speech grammars.

A page consists of dialogs (<form> and <menu> tags), event handlers, links (<link> tags), variable declarations (<var> tags), scripts (<script> tags), properties (<property> tags) document meta-data (<meta> tags).

When a page is initialized, properties are set, variables are declared and scripts are executed. Grammars in links and in dialogs with document scope are activated. Control is then passed to a dialog and optionally to a specific dialog item, depending on the URL. If the URL specifies only a file name, the first dialog is selected. If the URL specifies a dialog, then control passes to the named dialog. If this page was loaded due to a <goto> tag with the nextitem attribute set, this also determines the first item to be visited, otherwise the usual Form Interpretation Algorithm select phase criteria are used to select an item (the first item with the item variable undefined where the cond attribute is not false). If a root document is specified, it is also loaded and initialized before passing control to a dialog in the original page.

14 VoiceXML Reference: Attributes

accept

Exact or partial match.

Tags

<choice> <menu>

Values

Value	Description
"approximate"	Allow partial match.
"exact"	Require complete match.

Compatibility

VoiceXML 1.0 No, 2.0 Yes.

Description

When set to "approximate", specifies that part of a phrase may be matched.

When set to "exact", specifies that an entire phrase must be matched. This is used when two choices have sub-phrases in common, as in "business news" and "sports news", where "news" would be an ambiguous approximate match.

age
Age of TTS voice.

Tags
<voice>

Compatibility
VoiceXML 1.0 No, 2.0 Yes.

Description
Requests that the synthesized voice should sound like a person of the given age in years. It should be set to an integer value, for example age="50" or age="3".

VoiceXML browsers are not obligated to fully implement all speech markup attributes, however, a compliant browser must silently ignore any it does not support, and should make its best efforts to map any unimplemented mark-up to something similar that is supported.

alphabet

Phonetic alphabet.

Tags

<phoneme>

Compatibility

VoiceXML 1.0 No, 2.0 Yes.

Description

Specifies the name of the phonetic alphabet to use. May be set to "ipa", "worldbet", "xsampa" or another alphabet which is supported by the browser.

application
Root document.

Tags

<vxml>

Compatibility

VoiceXML 1.0 Yes, 2.0 Yes.

Description

Specifies the URL of the root document, which is a second VoiceXML page which is loaded in parallel with the page containing the current dialog. The root document is used to hold variables, links and dialogs which remain loaded even when the current page is swapped out (providing the new page references the same root document).

It is illegal for a page loaded as a root document to specify an application attribute.

archive

Location of resources needed by an object.

Tags

<object>

Compatibility

VoiceXML 1.0 Yes, 2.0 Yes.

Description

A space-separated list of URIs for archives containing needed by the object, which may include those resources specified by the classid and data attributes. URIs which are relative are interpreted relative to the codebase attribute.

audiobase
Based URL of vocabulary files.

Tags

<value>

Compatibility

VoiceXML 1.0 No, 2.0 Yes.

Description

If the browser supports playing a particular type of value by concatenating prompts, the value of this attribute may be used to specify options controlling how this is done.

It is envisaged that it will typically contain a URL which specifies a root directory for a set of vocabulary files recorded with a particular voice, however the interpretation of this value is browser-dependent.

bargein

Whether input interrupts prompt.

Tags

<prompt>

Values

Value	Description
"false"	Do not allow input to interrupt.
"true"	Allow input to interrupt.

Compatibility

VoiceXML 1.0 Yes, 2.0 Yes.

Description

If set to "true", user input will interrupt a prompt. If set to "false", user input will not interrupt a prompt. Bargein is typically disabled for error messages (to make sure the user doesn't continue unaware of the problem) and for paid sponsor content such as advertisements. Otherwise bargein is usually enabled since it allows an experienced user to move through the interface more quickly.

base

URL to use for relative links.

Tags

<vxml>

Compatibility

VoiceXML 1.0 Yes, 2.0 Yes.

Description

Specifies the base URL to use when interpreting relative URLs. Works in the same way as the <base> tag in HTML.

beep

Whether to play tone before recording.

Tags

<record>

Values

Value	Description
"false"	Do not play tone when recording starts.
"true"	Play a tone when recording starts.

Compatibility

VoiceXML 1.0 Yes, 2.0 Yes.

Description

Specifies whether or not to play a tone before starting to record.

bridge

Whether to make three-way call.

Tags

<transfer>

Values

Value	Description
"false"	Hang up when transfer completes.
"true"	Make a three-way call.

Compatibility

VoiceXML 1.0 Yes, 2.0 Yes.

Description

Specifies whether or not to make a three-way call when executing a <transfer> tag. When bridge is set to "true", the browser remains in the call with the user and the called party. When bridge is set to false, the browser disconnects from the call when the called party answers, leaving the user in a one-on-one conversation with that party.

caching

Whether to allow caching.

Tags

<audio> <choice> <goto> <grammar> <link> <object> <script>
<subdialog> <submit>

Values

Value	Description
"fast"	This resource may be cached.
"safe"	This resource may not be cached.

Compatibility

VoiceXML 1.0 Yes, 2.0 Yes.

Description

Set to "safe" to force a resource fetch every time the resource is used, or "fast" to enable use of a cache. The copy in the cache will expire in accordance with the HTTP response headers sent when the resource was fetched.

If no value is specified, the value from the innermost <property name="caching" .../> is used, or is set to "fast" if no such property has been set.

category
Preferred age category of TTS voice.

Tags

<voice>

Values

Value	Description
"adult"	Synthesized voice should sound adult.
"child"	Synthesized voice should sound child-like.
"elder"	Synthesized voice should sound like older person.
"teenager"	Synthesized voice should sound like teenager.

Compatibility

VoiceXML 1.0 No, 2.0 Yes.

Description

Specifies the preferred age category of the synthesized voice.

VoiceXML browsers are not obligated to fully implement all speech markup tags, however, a compliant browser must silently ignore any tags it does not support, and should make its best efforts to map any unimplemented mark-up to something similar that is supported.

charset

Character encoding.

Tags

<script>

Compatibility

VoiceXML 1.0 Yes, 2.0 Yes.

Description

Specifies the character encoding of the script, for example charset="iso-8859-1". For typical files recorded in ASCII or Unicode character sets the browser will be able to detect the character set automatically.

class
Kind of value.

Tags
<value>

Compatibility
VoiceXML 1.0 Yes, 2.0 Yes.

Description
Specifies the data type of the variable. May take the same values as the type attribute of the <say-as> tag.

classid
Location of object.

Tags
<object>

Compatibility
VoiceXML 1.0 Yes, 2.0 Yes.

Description
This attribute is a URL giving the location of the object's implementation. It might be a Java class file, ActiveX server or other type of object. The syntax of this attribute is browser-dependent.

codebase

Base URL to use for object resources.

Tags

<object>

Compatibility

VoiceXML 1.0 Yes, 2.0 Yes.

Description

The base path used to resolve relative URLs specified by the classid, data, and archive attributes. It defaults to the base URL of the current document as specified by the base attribute of the <vxml> tag or, if not specified, as deduced by the browser from the URL of the current page.

codetype

Context type of object data.

Tags

<object>

Compatibility

VoiceXML 1.0 Yes, 2.0 Yes.

Description

The MIME type or content type of the data expected when downloading
the object specified by the classid attribute. When absent it defaults to
the value of the type attribute.

cond

Boolean ECMAScript expression.

Tags

<block> <catch> <elseif> <error> <field> <help> <if> <initial> <noinput> <nomatch> <object> <prompt> <record> <subdialog> <transfer>

Compatibility

VoiceXML 1.0 Yes, 2.0 Yes.

Description

The cond attribute specifies an expression called a guard condition which is evaluated to give a boolean true or false value. When the value is false, the tag will be ignored or skipped. If not present, the guard condition is always considered to be true.

connecttimeout
Max time before reporting noanswer.

Tags

<transfer>

Compatibility

VoiceXML 1.0 Yes, 2.0 Yes.

Description

The length of time to wait before reporting no answer after initiating a transfer. The time is given as a numerical value with optional decimals followed by "s" for seconds or "ms" for milliseconds, for example "2.5s" or "2500ms".

content
Value of meta-data property.

Tags
<meta>

Compatibility
VoiceXML 1.0 Yes, 2.0 Yes.

Description
The value of a document meta-data property.

contour
Pitch contour for TTS.

Tags

<prosody>

Compatibility

VoiceXML 1.0 No, 2.0 Yes.

Description

Specifies how the pitch of the generated voice varies over time.

The contour takes precedence over both the pitch and range attributes. It is specified by a series of (position, pitch) pairs. The position is specified as a percentage of total time needed to speak the text, it is specified as a numerical value with optional decimal point followed by a percentage sign "%". The pitch can be specified using any value that would be valid for the pitch attribute. For example contour="(0%,+40%)(50%,-80%)".

count

Repeat count for prompt or handler.

Tags

<catch> <error> <help> <noinput> <nomatch> <prompt>

Compatibility

VoiceXML 1.0 Yes, 2.0 Yes.

Description

For prompts, specifies a minimum prompt count for the prompt to be played. The prompt counter is set to zero when a dialog is initialized and is increased by one each time the user is prompted for a given item. The prompt counter can be set to zero by using the <clear> tag.

The child prompt with the highest count attribute less than or equal to the prompt counter is used. If a prompt has no count attribute, a count of "1" is assumed.

For event handlers in a dialog or dialog item, specifies a minimum event count for the handler to be eligible to handle an event. The event counter is set to zero when the dialog is initialized and increased by one each time the event is triggered within that dialog or dialog item. If an event is thrown within a <filled> item, the form-level event counter is incremented.

data
Object data.

Tags

<object>

Compatibility

VoiceXML 1.0 Yes, 2.0 Yes.

Description

The URL of the object's serialized data (as opposed to the classid attribute, which specifies the location of the implementation). If it is a relative URL, it is interpreted relative to the codebase attribute.

dest

Destination URL (e.g. phone number) for transfer.

Tags

<transfer>

Compatibility

VoiceXML 1.0 Yes, 2.0 Yes.

Description

The destination URL for the transfer. Could for example specify a phone number using the "phone://" URL scheme, as in dest="phone://8005551212".

destexpr

ECMAScript expression for destination URL (e.g. phone number).

Tags

<transfer>

Compatibility

VoiceXML 1.0 Yes, 2.0 Yes.

Description

An ECMAScript expression which is evaluated to give the target URL for the transfer, e.g. a phone number to dial. See also the dest attribute.

dtmf

Specify touch-tone for choice.

Tags

<choice> <menu> <option>

Values

Value	Description
Text	Touch-tone digit for this choice.
"false"	Do not assign touch-tone digit(s).
"true"	Assign touch-tone digit(s).

Compatibility

VoiceXML 1.0 Yes, 2.0 Yes.

Description

For an <option> or <choice> tag, specifies a touch-tone digit to select the choice.

For a <menu> tag, is set to "true" or "false" to indicate whether touch-tone selections should be assigned automatically to choices.

dtmfterm
Whether to allow touch-tone interruption.

Tags

<record>

Values

Value	Description
"false"	Touch-tone does not terminate recording.
"true"	Touch-tone terminates recording.

Compatibility

VoiceXML 1.0 Yes, 2.0 Yes.

Description

Indicates whether a touch-tone digit should terminate recording. For fine control, set to "false" and specify a touch-tone grammar, i.e. a grammar with mode="dtmf".

duration

Desired time to speak text.

Tags

<prosody>

Compatibility

VoiceXML 1.0 No, 2.0 Yes.

Description

Specifies the total time it should take to speak the text enclosed by the tag. The value is a numerical value with optional decimals followed by "s" for seconds or "ms" for milli-seconds, for example "4.1s" or "1200ms".

enctype
MIME encoding for submitted data.

Tags

<subdialog> <submit>

Compatibility

VoiceXML 1.0 Yes, 2.0 Yes.

Description

MIME encoding to use for the submitted data. VoiceXML browsers must support "application/x-www-form-urlencoded", which uses URL encoding (either in the URL itself or in the entity body), and "multipart/form-data", which specifies a multi-part form (as in multiple MIME attachments). Browsers may also support other formats.

event

Name of event.

Tags

<catch> <choice> <link> <return> <throw>

Compatibility

VoiceXML 1.0 Yes, 2.0 Yes.

Description

Specifies the name of an event.

expr
ECMAScript expression.

Tags

<assign> <audio> <block> <choice> <exit> <field> <goto> <initial>
<link> <object> <param> <record> <subdialog> <submit> <transfer>
<value> <var>

Compatibility

VoiceXML 1.0 Yes, 2.0 Yes.

Description

Specifies an expression to evaluate. Depending on the tag, the
expression is used to set a variable, to set a URL or for other purposes.

expritem

ECMAScript expression which gives item name.

Tags

<goto>

Compatibility

VoiceXML 1.0 Yes, 2.0 Yes.

Description

An ECMAScript expression which evaluates to the name of a dialog item.

fetchaudio

URL of "Hourglass" sound.

Tags

<choice> <goto> <link> <subdialog> <submit>

Compatibility

VoiceXML 1.0 Yes, 2.0 Yes.

Description

The URL of the sound to play while fetching a VoiceXML page (the "hourglass sound"). If not specified, the innermost <property name="fetchaudio" .../> is used, or if that property is not set, no audio is played during the fetch.

fetchhint

Option for resource read-ahead.

Tags

<audio> <choice> <goto> <grammar> <link> <object> <script> <subdialog> <submit>

Values

Value	Description
"prefetch"	Allow read-ahead of resource.
"safe"	Get resource only when needed.
"stream"	Begin processing resource immediately.

Compatibility

VoiceXML 1.0 Yes, 2.0 Yes.

Description

Specifies when a resource may be fetched and when to begin using it.

If set to "prefetch", the browser may fetch the resource at any time after the page is loaded.

If set to "safe", the browser must fetch the resource at the time it is referenced.

If set to "stream", the browser must fetch the resource at the time it is referenced and should begin using it without waiting for the fetch to complete.

fetchtimeout

Max time to wait before triggering error.badfetch event.

Tags

<audio> <choice> <goto> <grammar> <link> <object> <script>
<subdialog> <submit>

Compatibility

VoiceXML 1.0 Yes, 2.0 Yes.

Description

The length of time to wait before throwing an error.badfetch event.

finalsilence

End recording when this much silence.

Tags

<record>

Compatibility

VoiceXML 1.0 Yes, 2.0 Yes.

Description

Specifies the length of silence which will terminate a recording. Specified as a numerical value with optional decimals followed by "s" for seconds or "ms" for milliseconds, for example "2.5" or "3000ms".

gender
Gender of TTS voice.

Tags

<voice>

Values

Value	Description
"female"	Request female TTS voice.
"male"	Request male TTS voice.
"neutral"	Request gender-neutral TTS voice.

Compatibility

VoiceXML 1.0 No, 2.0 Yes.

Description

Specifies the requested gender of the synthesized voice.

VoiceXML browsers are not obligated to fully implement all speech markup attributes, however, a compliant browser must silently ignore any it does not support, and should make its best efforts to map any unimplemented mark-up to something similar that is supported.

http-equiv

Name of an HTTP response header field.

Tags

<meta>

Compatibility

VoiceXML 1.0 Yes, 2.0 Yes.

Description

Specifies the HTTP response header field name, for example "Expires".

id

Name a tag for later reference in a URL.

Tags

<form> <menu> <rule>

Compatibility

VoiceXML 1.0 Yes, 2.0 Yes.

Description

Names a particular tag so that it can be referenced in a URL fragment by using the "URL#anchorname" syntax.

import

Alias name for grammar rule.

Tags

<ruleref>

Compatibility

VoiceXML 1.0 No, 2.0 Yes.

Description

See description of the <import> tag.

lang

Obsolete. *Language and locale type.*

Tags

<vxml>

Compatibility

VoiceXML 1.0 Yes, 2.0 No.

Description

Specifies the language and locale type. Replaced by xml:lang in VoiceXML 2.0.

level

TTS voice level.

Tags

<emphasis>

Values

Value	Description
"moderate"	TTS voice level.
"none"	TTS voice level.
"reduced"	TTS voice level.
"strong"	TTS voice level.

Compatibility

VoiceXML 1.0 Yes, 2.0 Yes.

Description

Requests a given level (volume) of the synthesized voice.

VoiceXML browsers are not obligated to fully implement all speech markup attributes, however, a compliant browser must silently ignore any it does not support, and should make its best efforts to map any unimplemented mark-up to something similar that is supported.

lexicon

Pronunciation database.

Tags

<token>

Compatibility

VoiceXML 1.0 No, 2.0 Yes.

Description

Specifies a lexicon to use in interpreting a token. The set of supported lexicons, if any, is browser-dependent.

maxtime

Max duration of the recording or transfer.

Tags

<record> <transfer>

Compatibility

VoiceXML 1.0 Yes, 2.0 Yes.

Description

Maximum duration of the recording or maximum time to allow for the transfer before reporting a "noanswer" result. Specified as a numerical value with optional decimals followed by "s" for seconds or "ms" for milliseconds, for example "120s" or "60000ms".

method

HTTP command name.

Tags

<subdialog> <submit>

Values

Value	Description
"get"	Use HTTP get method.
"post"	Use HTTP post method.

Compatibility

VoiceXML 1.0 Yes, 2.0 Yes.

Description

Specifies which HTTP method to use to submit the data.

modal

Whether to enable outside grammars.

Tags

<field> <record>

Values

Value	Description
"false"	Enable higher-level grammars that are in scope.
"true"	Disable higher-level grammars that are in scope.

Compatibility

VoiceXML 1.0 Yes, 2.0 Yes.

Description

If set to "true", specifies that all grammars outside the current item should be disabled while the item is being visited, even if those grammars are in scope.

mode
Miscellaneous mode specifications.

Tags

<filled> <grammar> <value>

Values

Value	Description
"all"	All fields.
"any"	Any field.
"dtmf"	Touch-tone input.
"recorded"	Request output by concatenated pre-recorded vocabulary.
"speech"	Speech recognition input.
"tts"	Output by TTS.

Compatibility

VoiceXML 1.0 Yes, 2.0 Yes.

Description

The meaning of and allowed values for this attribute depends on the tag.

For <value>, mode is set to "recorded" to request that the browser use concatenated pre-recorded sound files to play a value rather than synthesized speech, the default is "tts" which means to use speech synthesis.

For <grammar>, the mode specifies whether input is by speech recognition ("speech") or by touch-tones ("dtmf").

For <filled>, mode="any" specifies that the <filled> executable content is triggered when any one of the fields name in its namelist attribute are filled in by user input, mode="all" specifies that all fields must be filled in.

msecs
Obsolete. *Number of milliseconds.*

Tags
<break>

Compatibility
VoiceXML 1.0 Yes, 2.0 No.

Description
Obsolete. Specifies a length of time in milliseconds as a numerical value. Not supported in VoiceXML 2.0.

name
Name of an element.

Tags

<assign> <block> <field> <import> <initial> <mark> <meta> <object>
<param> <property> <record> <subdialog> <transfer> <var> <voice>

Compatibility

VoiceXML 1.0 Yes, 2.0 Yes.

Description

Specifies the name of an item, variable or parameter.

namelist

List of variable names.

Tags

<clear> <exit> <filled> <return> <subdialog> <submit>

Compatibility

VoiceXML 1.0 Yes, 2.0 Yes.

Description

A whitespace-separated list of variable or field names.

next
URL of next page and/or dialog.

Tags
<choice> <goto> <link> <submit>

Compatibility
VoiceXML 1.0 Yes, 2.0 Yes.

Description
The URL of the next page to load.

nextitem

First item to visit in next dialog.

Tags

<goto>

Compatibility

VoiceXML 1.0 Yes, 2.0 Yes.

Description

The name of the item to visit first in the next dialog.

number

Repeat count for a rule.

Tags

<count>

Compatibility

VoiceXML 1.0 No, 2.0 Yes.

Description

Specifies how many times the enclosed grammar rule must be found in the user input. Can be set to "optional" or "?" to indicate zero or one times, "0+" to indicate zero or more times, or "1+" to indicate one or more times.

ph

Phoneme string.

Tags

<phoneme>

Compatibility

VoiceXML 1.0 No, 2.0 Yes.

Description

Specifies the phoneme string using the alphabet given by the alphabet attribute.

phon

Obsolete. *Phoneme string.*

Tags

<sayas>

Compatibility

VoiceXML 1.0 Yes, 2.0 No.

Description

Obsolete. Replace by the ph attribute in VoiceXML 2.0.

pitch
Pitch of the TTS voice.

Tags

<prosody>

Compatibility

VoiceXML 1.0 Yes, 2.0 Yes.

Description

Requests the pitch (average frequency) in the synthesized voice. Specified as a numerical value for a frequency (in Hz), e.g. "150", a relative change e.g. "+10%", or one of "high", "medium", "low", "default".

Relative changes to values are specified as signed numbers with optional decimals, for example "+12.5". The change can be expressed as a percentage, e.g. "-10%" or as a number of musical semi-tones, e.g. "+3.5st".

The typical baseline (average) pitch for a female voice is between 140 Hz and 280 Hz, with a pitch range of 80 Hz or more. Male voices are typically lower with a baseline of 70 to 140 Hz and a range of 40 to 80Hz.

VoiceXML browsers are not obligated to fully implement all speech markup attributes, however, a compliant browser must silently ignore any it does not support, and should make its best efforts to map any unimplemented mark-up to something similar that is supported.

range
Range of the TTS voice.

Tags

<prosody>

Compatibility

VoiceXML 1.0 Yes, 2.0 Yes.

Description

Requests a range between the maximum and minimum pitch in the synthesized voice. See pitch.

rate

Speaking rate.

Tags

<prosody>

Compatibility

VoiceXML 1.0 Yes, 2.0 Yes.

Description

Requests a speaking rate of synthesized speech in words per minute. The value is give as a absolute number of words per minute, relative change, or one of "fast", "medium", "slow", "default".

Relative changes to values are specified as signed numbers with optional decimals, for example "+12.5". The change can be expressed as a percentage, e.g. "-10%".

VoiceXML browsers are not obligated to fully implement all speech markup tags, however, a compliant browser must silently ignore any tags it does not support, and should make its best efforts to map any unimplemented mark-up to something similar that is supported.

recsrc
Obsolete. *Base URL for audio vocabulary.*

Tags
<value>

Compatibility
VoiceXML 1.0 Yes, 2.0 No.

Description
Obsolete. Specifies the method to use when concatenating audio.
Replaced by audiobase in VoiceXML 2.0.

root

Root rule in grammar.

Tags

<grammar>

Compatibility

VoiceXML 1.0 No, 2.0 Yes.

Description

Specifies the root rule for the grammar. The rule must have public scope. If not specified, the root rule is formed by making a choice of all the top-level rules defined in the grammar.

scope

Range where an element may be used.

Tags

<form> <grammar> <menu> <rule>

Values

Value	Description
"dialog"	Scope within current dialog.
"document"	Scope within current document.
"private"	Private grammar rule.
"public"	Public grammar rule.

Compatibility

VoiceXML 1.0 Yes, 2.0 Yes.

Description

Specifies the scope of the tag, in other words, the range in which the name of that tag can be referenced or, in the case of a grammar, when the grammar is active.

For a grammar rule, the scope can be "private", meaning that the rule can only be referenced within the grammar, or "public", meaning that the rule can be referenced inside or outside the grammar.

For a <grammar> tag which is a child of a form, the scope may be set to "dialog" or "document". Dialog scope means that the grammar is active when the form is being visited, document scope means that the grammar is active for the entire time that the current page is loaded. It is illegal to specify a scope for a grammar which is not the child of a form. Field grammars are active when the field is being visited, link grammars have the scope of the parent of the link. For example, if the link is a child of a form, the grammar is in scope when that form is being visited.

For <form> and <menu> tags, the scope may be set to "dialog" or "document" which determines the default scope of a dialog-level grammar.

size
Length of pause.

Tags
<break>

Values

Value	Description
"large"	Long pause.
"medium"	Medium pause.
"none"	No pause.
"small"	Small pause.

Compatibility
VoiceXML 1.0 Yes, 2.0 Yes.

Description
Specifies the length of a pause in synthesized speech.

slot
Named position within a grammar.

Tags

<field>

Compatibility

VoiceXML 1.0 Yes, 2.0 Yes.

Description

Specifies a slot name for the field. This is used for grammar types which support defining slot/value pairs. By default, the name of the slot is expected to be the same as the name of the field, this allows a different slot name to be used.

special
Name of a pre-defined rule.

Tags

<ruleref>

Compatibility

VoiceXML 1.0 No, 2.0 Yes.

Description

Set to one of the special built-in rules "#VOID", "#GARBAGE" or "#NULL".

src
URL of the resource.

Tags
<audio> <grammar> <script> <subdialog>

Compatibility
VoiceXML 1.0 Yes, 2.0 Yes.

Description
Specifies the URL.

sub
Substitute text.

Tags
<say-as>

Compatibility
VoiceXML 1.0 Yes, 2.0 Yes.

Description
Substitute text. See <say-as> tag.

tag
Name of grammar item.

Tags

<count> <item> <one-of> <ruleref>

Compatibility

VoiceXML 1.0 No, 2.0 Yes.

Description

Specifies a syntactic tag. This is a preliminary proposal in the VoiceXML 2.0 draft specification which has not been finalized.

time
Duration.

Tags

<break>

Compatibility

VoiceXML 1.0 No, 2.0 Yes.

Description

Specifies the length of time to pause in the synthesized speech. Given as a numerical value with optional decimals followed by "s" for seconds or "ms" for milliseconds, for example "1.5s" or "1500ms".

timeout
Max time to wait before triggering noinput event.

Tags
<prompt>

Compatibility
VoiceXML 1.0 Yes, 2.0 Yes.

Description
Maximum time to wait for user input following a prompt before triggering a noinput event. Given as a numerical value with optional decimals followed by "s" for seconds or "ms" for milliseconds, for example "1.5s" or "1500ms".

type
How to interpret text to be spoken.

Tags

<field> <grammar> <object> <param> <record> <say-as>

Values

Value	Description
Text	Content (MIME) type.
"acronym"	Speak acronym as letters, not word.
"address"	Speak as address.
"currency"	Money amount.
"date:d"	Day of month.
"date:dmy"	Day, month and year.
"date:m"	Month.
"date:md"	Month and day.
"date:mdy"	Month, day and year.
"date:my"	Month and year.
"date:y"	Year.
"date:ym"	Year and month.
"date:ymd"	Year, month and day.
"duration:h"	Hours.
"duration:hm"	Hours and minutes.
"duration:hms"	Hours, minutes and seconds.
"duration:m"	Minutes.
"duration:ms"	Minutes and seconds.
"duration:s"	Seconds.
"measure"	Measurement.
"name"	Speak person's name.
"net"	URL.
"number:digits"	Digit string one, two, three...
"number:ordinal"	First, second, third....
"telephone"	Phone number.
"time:h"	Hours.
"time:hm"	Hours and minutes.
"time:hms"	Hours, minutes and seconds.

Compatibility

VoiceXML 1.0 Yes, 2.0 Yes.

Description

The meaning of this attribute depends on the tag: for <object>, <grammar>, <record> and <param>, it is a content type (MIME type), for <field> and <say-as>, it is a data type.

For an <object> tag, it specifies the content type of the data specified by the data attribute.

For a <param> tag in an <object> tag where the valuetype attribute is set to "ref", it specifies the content type of the resource.

For a <grammar> tag, it specifies the content type of the grammar. The XML form of SRGF is specified as "application/grammar+xml" and the ABNF form as "application/grammar". The browser may be able to determine the content type from the HTTP response header or from the filename extension. Proposed standard extensions for SRGF files are ".grxml" for the XML form and ".grm" for the ABNF form.

For a <record> tag, it specifies the content type of the audio file to record.

For a <field> tag, it specifies a built-in type for the field, one of "boolean", "date", "digits", "currency", "number", "phone" or "time".

For a <say-as> tag, it specifies the data type to use in interpreting the value, one of: "acronym", "number", "number:ordinal", "number:digits", "telephone", "date", "date:dmy", "date:mdy", "date:ymd", "date:ym", "date:my", "date:md", "date:y", "date:m", "date:d", "time", "time:hms", "time:hm", "time:h", "duration", "duration:hms", "duration:hm", "duration:ms", "duration:h", "duration:m", "duration:s", "currency", "measure", "name", "net", or "address". Note that "acronym", perhaps counter-intuitively, means to speak a group of letters as a word, as in "DEC" or "NASA", rather than the default, which is to spell out the letters, as in "IBM" or "USA".

uri
URL of resource.

Tags
<import> <ruleref>

Compatibility
VoiceXML 1.0 No, 2.0 Yes.

Description
Specifies a URL. Technically, a Uniform Resource Indicator, which is a superset of a Uniform Resource Locator. We use the term "URL" in this reference since we believe it is more familiar to readers and is for most practical purposes the same, at least with today's Web.

value
Value to assign.

Tags
<option> <param> <property>

Compatibility
VoiceXML 1.0 Yes, 2.0 Yes.

Description
For <property> or <param>, the value to assign to the property or parameter named by the name attribute.

In the case of <option>, it is the value to assign to the field variable if that option is selected.

valuetype
Whether data is a URL.

Tags

<param>

Values

Value	Description
"data"	Value itself is data.
"ref"	Value is a URL pointing to data.

Compatibility

VoiceXML 1.0 Yes, 2.0 Yes.

Description

If the parameter is being passed to an object, the valuetype attribute may be set to "ref" or "data" to indicate if the value in the data attribute is an URL which points to the value ("ref"), or is itself the value to use ("data").

variant
Browser-dependent voice variant.

Tags

<voice>

Compatibility

VoiceXML 1.0 No, 2.0 Yes.

Description

Set to an integer value to select a browser-dependent variant of the synthesized voice. For example, <voice gender="female" variant="2"> might select the second woman's voice supported by the synthesizer.

version

Language version.

Tags

<grammar> <vxml>

Compatibility

VoiceXML 1.0 Yes, 2.0 Yes.

Description

Specifies the VoiceXML or grammar language version.

vol

Obsolete. *Volume level of TTS voice.*

Tags

<pros>

Compatibility

VoiceXML 1.0 Yes, 2.0 No.

Description

Specifies the volume of the synthesized speech. Obsolete, replace by volume in VoiceXML 2.0.

volume
Volume level of TTS voice.

Tags

<prosody>

Compatibility

VoiceXML 1.0 No, 2.0 Yes.

Description

Requests a volume level for the synthesized voice. Given as a number with optional decimals in the range "0.0" (silent) to 100.0 (loudest), a relative change, or values "silent", "soft", "medium", "loud" or "default".

Relative changes to values are specified as signed numbers with optional decimals, for example "+12.5". The change can be expressed as a percentage, e.g. "-10%".

VoiceXML browsers are not obligated to fully implement all speech markup tags, however, a compliant browser must silently ignore any tags it does not support, and should make its best efforts to map any unimplemented mark-up to something similar that is supported.

weight
Relative probability of a choice.

Tags

<item>

Compatibility

VoiceXML 1.0 No, 2.0 Yes.

Description

Specifies the relative weight to give to an item when attempting to match a grammar rule. It is ignored unless it is an attribute of an <item> within a <one-of>.

The weight should indicate the estimated probability of this item occurring versus the other items. Weights are numerical values with optional decimals and must be zero or greater. In the absence of weights or if the weights are not properly specified (e.g. one or more missing) then the choices are assumed to be equally likely. There are no particular units, the sum of all the weights within a given <one-of> is considered equivalent to 100% likelihood. So if the weights for a choice with three items were "0.2", "0.2" and "0.6", this would represent 20%, 20% and 60%. If the weights were "3", "3" and "9", this would have exactly the same effect.

xml:lang

Language.

Tags

<count> <grammar> <item> <one-of> <p> <paragraph> <prompt>
<ruleref> <s> <sentence> <token> <vxml>

Compatibility

VoiceXML 1.0 No, 2.0 Yes.

Description

Specifies the language and optionally a locale. Child tags inherit the
xml:lang value recursively from their parents. An inner tag may override
the value specified by a parent.

The format of the value is a two-letter language code as specified by ISO
639, such as "en" for English, followed optionally by a hyphen and a sub-
specifier, which may be an ISO 3166 country code, such as "US" for the
United States or "GB" for Great Britain. A typical value is "en-US" for
English (US).

The browser defines a default value. If the browser encounters an
xml:lang attribute that it does not support, it throws an error event.

15 VoiceXML Reference: Events, Properties and Variables

15.1 Events

The following events are pre-defined by the VoiceXML language.

Event	Description
cancel	User input matched the built-in cancel grammar.
error.badfetch	A resource is required but the browser was unable to fetch it from the specified URL.
error.badfetch.http.nnn	HTTP error nnn occurred, for example, 404 for Page Not Found.
error.noauthorization	"Access denied" error, user has tried to do something without sufficient permission.
error.semantic	A semantic error was found in the VoiceXML page, for example an undefined variable was referenced, divide by zero, root document <vxml> tag has an application attribute, etc.
error.unsupported.format	Browser does not support the resource format, e.g. sound file type, grammar type etc.
error.unsupported.tag	The browser does not support the given tag. E.g., if the browser does not support <subdialog>, it would throw error.unsupported.subdialog.
exit	User input matched the built-in exit grammar.
help	User input matched the built-in help grammar.
noinput	No user input was expected but none was detected.
nomatch	User input did not match the

Event	Description
	grammar for the current form or field.
telephone.disconnect. hangup	The user hung up or a <disconnect> was executed.
telephone.disconnect. transfer	The browser hung up as a result of an unconditional transfer.

15.2   Event handlers

The following default event handlers are built into a VoiceXML browser.

Event	Default handler
cancel	Do nothing (continue to interpret the current dialog).
error	Message, <exit>.
exit	<exit>.
help	Message, <reprompt>.
noinput	<reprompt>.
nomatch	Message, <reprompt>.
telephone. disconnect	<exit>.
Other	Message, <exit>.

"Message" indicates that the browser plays a message, the message content is browser-dependent.

15.3 Shadow variable properties

Some types of dialog item have a shadow variable which is named $itemname, where itemname is specified by the name attribute. Following are the properties of the shadow variable.

Item	Property	Description
<field>	confidence	Estimate of the probability of a correct match by the speech recognizer, in the range 0.0 to 1.0.
<field>	inputmode	"dtmf", "speech" or "voice" to indicate whether touch-tone or speech input filled the field. "voice" and "speech" are synonymous, same as the mode attribute of the <grammar> tag if applicable.
<field>	utterance	Text form of the recognizer match. Spelling and punctuation (e.g. whether numbers are digits or words) will be browser-dependent.
<record>	duration	Duration of the recording in milliseconds.
<record>	size	Size of the recording in bytes. This is the raw audio data, it does not include any file header.
<record>	termchar	The touch-tone digit which terminated the recording or the ECMAScript special value null if the recording terminated for some other reason.
<transfer>	duration	Time from starting to dial until the call was answered in seconds.

15.4 Properties

Following are the properties which are defined by the VoiceXML language. They are set using the <property> tag. There may be other, browser-dependent properties.

Property	Description
audiofetchhint	Default value for the fetchhint attribute of <audio> tags. Default is "prefetch".
bargein	Default value for bargein attribute. Default is "true".
caching	Default value for caching attribute. Default is "fast".
completetimeout	Silence time used by the recognizer to determine end of utterance, as a numerical value followed by "s" for seconds or "ms" for milliseconds, when a complete match to a grammar has been found. Typical values are from "0.3s" to "1.0s".
confidencelevel	The speech recognition confidence level, a numerical value in the range of "0.0" (zero probability of match) to "1.0" (certain match). A nomatch event is thrown when the reported confidence level is less. Default is "0.5".
documentfetchhint	Default value for the fetchhint attribute of tags which fetch VoiceXML pages. Default is "safe".
fetchaudio	Default value for fetchaudio attribute. Default is no audio.
fetchaudiodelay	Time to wait after starting fetch before beginning to play fetchaudio as a numerical value followed by "s" for seconds or "ms" for milliseconds. Default is browser-dependent.

Property	Description
fetchaudiominimum	Minimum time to play fetchaudio as a numerical value followed by "s" for seconds or "ms" for milliseconds. Once this time has expired, the fetchaudio will be interrupted as soon as the fetch has completed. Default is browser-dependent.
fetchtimeout	Default value for fetchtimeout attribute. Default is browser-dependent.
grammarfetchhint	Default value for the fetchhint attribute of <grammar> tags which specify external grammars. Default is "prefetch".
incompletetimeout	As for completetimeout, except that input matches only a prefix of a grammar. A nomatch event will be triggered when this timeout expires.
inputmodes	Input modes to enable, as a white-space separate list including "dtmf" and/or one of "voice" or "speech", which are synonyms. Default is browser-dependent.
interdigittimeout	Length of time between touch-tone digits used to determine end of input, as a numerical value followed by "s" for seconds or "ms" for milliseconds. Default is browser-dependent.
objectfetchhint	Default value for the fetchhint attribute of <object> tags. Default is "prefetch".
scriptfetchhint	Default value for the fetchhint attribute of <script> tags which specify external scripts. Default is "prefetch".

Property	Description
sensitivity	The speech recognizer threshold for accepting quiet input in the range "0.0" (require high amplitude input with no background noise) to "1.0" (accept low amplitude input with high background noise). Default is "0.5".
speedvsaccuracy	Hint to the speech recognizer on trade-off for recognition speed vs. accuracy (more accuracy generally requires more processing time) from "0.0" (fastest) to "1.0" (most accurate). Default is "0.5".
termchar	A touch-tone digit which terminates input, default is "#".
termtimeout	Length of time following the last touch-tone detected used to determine end of input, , as a numerical value followed by "s" for seconds or "ms" for milliseconds. If "0s" (the default), then the recognized value is returned immediately after the last DTMF allowed by the grammar.
timeout	Default value for timeout attribute. Default is browser-dependent.
universals	Universal grammars to enable, as a white-space separate list including "cancel", "exit" and/or "help", or one of "all" (to enable all) or "none" (to disable all). Default is "all".

15.5 Session variables

Following are the session variables defined by the browser. They can be referenced using the variable name alone, however it is recommended that you use the safer style session.variablename, for example session.telephone.ani. If the value is not known or not applicable to the current call, the value is set to undefined.

Session variable	Description
telephone.ani	ANI digits.
telephone.dnis	DNIS digits.
telephone.iidigits	Information Indicator Digits. Information about the originating line (e.g. payphone, cellular service, special operator handling, prison) of the caller.
telephone.rdnis	The number from which a call diversion or transfer was invoked.
telephone.redirect_reason	Will be set to one of the following strings if rdnis is defined: "unknown", "user-busy", "no-answer", "unavailable", "unconditional", "time-of-day", "do-not-disturb", "deflection", "follow-me", "out-of-service", "away".
telephone.uui	User to User Information. A user-defined field in ISDN.

16 VoiceXML Reference: Changes from Version 1.0 to 2.0

16.1 VoiceXML 1.0

The VoiceXML 1.0 standard was published on March 7th, 2000 by the VoiceXML Forum. In this section we describe the differences between this version 1.0 standard and the December 14th, 2000 draft of the proposed W3C VoiceXML 2.0 standard.

Changes are summarized in the following table.

Change from 1.0 to 2.0	Description
SRGF added to DTD	These tags were added in 2.0: <count> <example> <import> <item> <one-of> <rule> <ruleref> <token>
SSML added to DTD	These tags were added in 2.0: <emphasis> <mark> <p> <paragraph> <phoneme> <prosody> <s> <say-as> <sentence> <voice>
Debug logging added	<log> tag was added.
Tags re-named.	These tags were re-named: <div> to <s> or <p> <emp> to <emphasis> <ph> to <phoneme> <pros> to <prosody> <sayas> to <say-as>
Attributes re-named.	These attributes were re-named: lang to xml:lang msecs to time phon to ph recsrc to audiobase vol to volume
Attributes added.	These attributes were added: accept age alphabet category contour duration gender import lexicon number root special tag uri variant weight

Change from 1.0 to 2.0	Description

Defaults attributes removed from DTD.	In VoiceXML 1.0, the following defaults were applied through the DTD:
	  caching="fast"
	  fetchhint="safe"
	  baregin="true"
	
	This was inappropriate because the defaults could in fact be overridden by <property>, in 2.0 these defaults were removed from the DTD.

Encoding type default corrected.	In VoiceXML 1.0, the default for the enctype attribute was defined by the DTD to be:
	  "application/x-www-formurlencoded"
	
	This was a typographical error. In VoiceXML 2.0, it has been corrected to:
	  "application/x-www-form-urlencoded".

Other changes	The expr attribute was added to <audio>. The type attribute of <field> was changed from a fixed list of values "boolean", "currency" ... to CDATA. This allows a browser to support additional, non-standard built-in types. The mode argument of <grammar> was added and a default value "1.0" for the version specified. In VoiceXML 1.0 there was no default for the size attribute of <break>, in 2.0 it is specified as "medium".

17 Glossary

μ-law	See mu-law.
ABNF	See Augmented Backus-Naur Form.
ACD	See Automatic Call Distributor.
Adaptive Differential Pulse Code Modulation	Compression scheme used for digital audio. It is a "lossy" scheme which tends to degrade the signal to some extent. Typically, ADPCM reduces the number of bits per sample from 8 to 4.
ADPCM	See Adaptive Differential Pulse Code Modulation.
A-law	Type of companding scheme used in the telephone network outside of North America.
analog	Represents signals by using a continuously variable physical quantity, such as the amount of current flowing.
ANI	See Automatic Number Identification.
API	See Applications Programming Interface.
Applications Programming Interface	A set of functions provided to a programming language for performing a logically related set of tasks.
ASR	See Automatic Speech Recognition.
associative arrays	A feature of a programming language where a data type other than an integer (typically a string) is used as an array subscript.
associativity	Property of an operator in a programming language which specifies whether it groups from left-to-right or right-to-left. E.g., addition is usually right-associative, meaning that a+b+c is interpreted as (a+b)+c, assignment is usually left-associative so that a=b=c is interpreted as a=(b=c).
audiotext	A type of voice application which offers pre-recorded information accessed via a fixed menu tree.
Augmented Backus-Naur Form	A particular variant of Backus-Naur Form.

auto-dialer	A device which dials out-bound calls (usually attempting many calls in parallel), when a call is completed it is patched through to a live operator. Used for tele-sales, polling etc.
Automatic Call Distributor	A switch which is designed for call centers, includes features for managing queues of in-bound callers who are on hold waiting to speak to a live agent
Automatic Number Identification	A service which provides the caller's number to the receiving equipment, similar to domestic Caller ID service. Usually, the number but not the subscriber name is provided.
Automatic Speech Recognition	Technology which converts digitized speech into text.
B channel	See bearer channel.
Backus-Naur Form	A text language used to specify the grammars of programming languages.
barge-in	A feature of speech recognition technology which allows the user to begin speaking before the prompt has finished, interrupting the prompt.
Basic Rate Interface	A type of ISDN interface that provides two 64 kbps channels for voice plus one channel for signaling.
battery voltage	Voltage applied to an analog phone line by a switch.
bearer channel	ISDN channel that carries voice. Also called B channel.
blind transfer	An attempt to transfer a telephone call to a new number which does not wait to determine the result of dialing the second call.
BNF	See Backus-Naur Form.
BRI	See Basic Rate Interface.

bridged

1. Calls are said to be "bridged" across a CT system when two callers are connected via the internal voice bus of that system.

2. "Bridged" transfer is used in VoiceXML to indicate that the browser does not hang up after the transfer, creating a three-way call with the user, browser and called party.

cache

Local storage used to improve performance by storing copies of resources fetched from the server and therefore reducing requests to a server for those resources.

cadence

1. The shape of speech caused by changes in pitch, without cadence, speech is monotonic.

2. The pattern of sound and silence used in signaling tones such as US busy tone ("beep...beep...beep").

call center

A group of live agents or operators who handle in-bound and/or out-bound calls in volume, typical examples would be technical support centers, catalogue sales representatives etc.

call completion

The process of connecting a call between two endpoints, a call is completed when both ends can talk to eachother.

call leg

Part of the path between two endpoints, either between an endpoint and a switch or between two switches.

call progress

The part of a call between dialing out and completing the call.

call progress analysis

An algorithm in automated telephony equipment that attempts to determine the result of an outbound call by monitoring call progress, the result would be "answered", "busy", "no answer" etc.

call transfer

An existing call is re-routed to a new endpoint.

computer-directed

Another term for machine-directed.

computer-initiative

Another term for machine-directed.

content	In general, refers to useful data (news, weather reports, sports scores, technical reference material, photographs...).
	In XML specifically, is another term for the value of an XML tag, that is, the text and/or sub-tags, if any, which appear between matching begin and end tags.
content provider	A vendor or system which supplies useful data (news, weather reports, sports scores, technical reference material, photographs...).
D channel	Type of ISDN channel which carries signaling data.
DNS	See Domain Name Service.
Domain Name Service	A standard Internet protocol which translates domain names such as intel.com into IP addresses.
Caller ID	A service on consumer phone lines which transmits the name and/or telephone number of the caller on an inbound call.
CAS	See Channel Associated Signaling.
Central Office	A telephone company switching station.
Channel Associated Signaling	A general tone for in-band tone signaling used on E1 trunks
clear-channel audio	A telephony channel which does not modify the transmitted digital audio (unlike robbed-bit T1, which makes slight, inaudible modifications to the audio to transmit signaling).
closing tag	The second of a pair *<tagname>* ... *</tagname>* in HTML or XML.
CO	See Central Office.
co-articulation	Where a repeated phoneme is pronounced only once in natural speech, for example "hot towel" is pronounced "hot owl".
collect phase	A step in the VoiceXML FIA where the browser waits for user input.

companding

A technique to improve the quality of audio represented by a given number of bits by giving finer detail at low amplitudes at the expense of coarse detail at high amplitudes (where the human ear is less sensitive).

Computer Telephony

Refers to any technology where a computer (other than an embedded system) is involved in processing a phone call.

Computer Telephony Integration

Refers to a configuration where an external computer is used to control a switch.

continuous recognition

Voice recognition which does not require the user to pause between words.

CT

See Computer Telephony.

CTI

See Computer Telephony Integration.

CTI link

The physical link between a CTI computer and a CTI switch.

customer care center

A call center providing customer service, such as a technical support line.

DAC

See Dialog Application Component.

DC

See Dialog Component.

decadic signaling

Another term for rotary dial or pulse signaling.

dial tone

The sound played by a switch when it is ready to receive digits.

Dialed Number Identification Service

A service which provides the number a caller dialed. This allows a single trunk to terminate several telephone numbers, which is used for example by hosting services.

dialog

An interaction between a computer and a user. In VoiceXML, specifically refers to a *<menu>* or *<form>*.

Dialog Application Component

Another term for a Dialog Component.

Dialog Component

A pre-compiled algorithm for a voice dialog.

Dialog Module	A proprietary term used by Speechworks to describe a Dialogic Component.
Digital Signal Processor	A specialized processor designed for digital signal analysis.
diphone	A sound constructed from a pair of phonemes, for example the word "me" is one diphone ("m" and "e").
direct-to-board API	An API where function calls are forwarded from the host PC directly to the firmware on the board without significant abstraction or a client/server layer.
DirectX	A Windows API for multimedia processing.
discrete recognition	A type of voice recognition technology which requires the user to pause between words.
disyllable	A pair of syllables, for example the word "expert" is a disyllable composed of the two syllables "ex-" and "-pert".
DNIS	See Dialed Number Identification Service.
DNS	See Domain Name Service.
document scope	A scope which includes the current VoiceXML page. See also scope.
Document Type Definition	A specification of legal XML syntax for a given type of document such as VoiceXML.
DS0	Digital Signal level 0. A single channel on a digital trunk carrying 64 kbps data (8 kHz 8 bit PCM with A-law or mu-law companding).
DS1	Digital Signal level 1. Synonym for T1.
DS3	Digitial Signal level 3. A single trunk which carries 576 DS0 channels (equivalent to 24 T1s or 18 E1s).
DSP	See Digital Signal Processor.
DTD	See Document Type Definition.
DTMF	See Dual Tone Multi Frequency.

dual tone	A tone composed of two pure frequencies, i.e. the combination of two sine waves of different periods.
Dual Tone Multi-Frequency	Also called touch-tones. A set of 16 dual tones used to represent the digits 0 - 9 plus #, *, A, B, C, D.
echo cancelation	The process of subtracting a prompt being played from the audio to be analyzed by a voice recognizer.
ECMA-262	The ECMA standard defining the exact version of ECMAScript used in VoiceXML.
ECMAScript	An industry-standard scripting language.
ECTF	Enterprise Computer Telephony Forum. An industry forum which develops computer telephony standards.
element	The technical term used for an XML tag. We prefer to use "tag" in this book.
empty tag	A tag which encloses nothing, i.e. contains no value, as in <empty/>.
end-pointing	The process of determining the start and end of the utterance which should be recognized in voice recognition.
entity	A named string of text in an XML document, can be used as a short-hand analogous to a #define'd macro in C.
Euro-ISDN	One type of E1 ISDN found in Europe.
event	A notification to an application of a given condition. A typical event would report when a caller hangs up.
Fast Fourier Transform	An algorithm which determines the frequency components of digitized audio.
feature extraction	The process used by a voice recognizer to assign numerical values to relevant features of the analyzed audio, such as the frequency spectrum and how it changes with time.
FIA	See Form Interpretation Algorithm.

flash-hook	A signal sent on an analog phone by depressing the hook switch for a short time.
form	A description in a computer language of an interaction with the user where values are collected by filling in fields.
Form Interpretation Algorithm	The algorithm used by a browser when interpreting a VoiceXML page.
Fourier analysis	A mathematical technique for splitting up a signal into its frequency components.
frame	In TDM, a frame contains exactly one time-slot from each channel.
G.723.1	A compression scheme in H.323 IP telephony which offers very low bit rates.
Gateway	Equipment which translates between two different types of network, for example between an IP network and the PSTN.
glare	An attempt to start an outbound call instead answered an inbound call.
grammar	A template which describes a set of strings. In VoiceXML, a grammar describes strings of spoken words.
H.323	A protocol for IP telephony.
headerless file	A file which contains only raw data without a header.
hook switch	The switch on an analog phone which is turned off and on by raising and lowering the handset from its cradle.
HTML	See Hyper Text Markup Language.
in-band	Signaling which uses the same channel that transmits voice.
Indexed Prompt File	A type of voice file that contains more than one recording.
in-line grammar	A grammar that is specified in-place in the same page where it is used (as opposed to an external

grammar, which is stored in a separate file).

Integrated Services Digital Network
A set of protocol standards for digital trunks.

Interactive Voice Response
A general term for automated telephony equipment which interacts with the caller using spoken prompts and accepts user input in the form of touch-tones or voice recognition.

International Phonetic Alphabet
A standard phonetic alphabet.

International Telecommunications Union
An international standards body.

Internet Telephony
IP telephony on the Internet.

interpreter context
The technical name given by the VoiceXML standards documents to the software which controls the session. We use the informal term "browser" to refer to both the interpreter and the interpreter context.

IP Telephony
A general term for using IP networks to carry voice calls.

IPA
See International Phonetic Alphabet.

IPF
See Indexed Prompt File.

ISO/IEC 16262
The ISO standard which is equivalent to ECMA-262.

ITU
See International Telecommunications Union.

IVR
See Interactive Voice Response.

Java Speech Grammar Format
A Java standard for voice grammars.

Java Speech Markup Language
A Java standard for text-to-speech markup.

Java Telephony API
A set of standard Java classes.

JavaScript
Netscape's name for ECMAScript.

jitter
Interruptions in a streamed media caused by varying delay in arrival times of received packets.

JScript	Microsoft's name for ECMAScript.
JSGF	See Java Speech Grammar Format.
JSML	See Java Speech Markup Language.
JTAPI	See Java Telephony API.
last mile	The phone line between a telephone company's switching station and a residential consumer. Also called the local loop when it is an analog line.
linear	A relationship between two variables x and y which can be expressed as $x=ay + b$ where a and b are fixed. When applied to digital audio samples, means that the size of the sample has a linear relationship with the amplitude of the sound.
listening	A device on a TDM bus or trunk is said to listen to a time-slot when it is extracting an audio channel from that time-slot.
local loop	The phone line between a telephone company's switching station and a residential consumer. Also called the last mile.
loop current	The current that flows on the local loop.
loop drop disconnect supervision	Signals hang-up of the other party by turning off battery voltage for a short time, resulting in a drop in loop current to zero.
loop start	An analog phone line where a call is initiated by going off-hook, which causes loop current to flow.
machine-directed	A conversation in which the computer prompts the user for input and the user responds. Typical examples are a login dialog which requires a user name and password to be entered, or a touch-tone menu where the computer asks the user to dial a digit.
machine-intiative	Another term for machine-directed.
Media Gateway	A type of Gateway whose main function is to convert between different types of media transport, for example between IP and PSTN.

menu

A user interface idiom where the computer presents a fixed series of choices to the user, who responds by selecting one of those choices.

message store-and-forward

A computer system which stores messages intended for a given user, and allows that user to retrieve those messages at a later time. E-mail and voice mail are the canonical store-and-forward systems.

MF

Multi Frequency. An alternative system of dual tones to DTMF, also called R1.

mixed-initiative

An interaction between a user and a computer which is sometimes user-driven and sometimes machine-driven. A typical example would be a voice command "I want to buy a ticket from New York to Boston" where the computer responds "What day do you want to fly?".

mouse hover

The operation of moving the mouse over a graphical user interface element and leaving it there for a short time. The interface may show a short help message or respond in some other way.

mu-law

A type of companding used by the phone network in North America.

NFAS

An ISDN feature which allows multiple trunks to share a single D channel, thereby increasing the number of B (voice) channels available.

Nyquist's Theorem

A mathematical theorem in digital audio which states that the maximum frequency which can be carries is half the sampling rate, so an 8 kHz channel cannot carry frequencies higher than 4 kHz.

offering

The technical term for a phone line which is presenting an incoming call, the colloquial term would be "ringing".

on-hook

The state where the handset of an analog telephone is in its cradle, setting the hook switch to the off position and preventing current from flowing.

opening tag

The first tag in a pair <tagname>...</tagname>.

ordinal number

A number expressing rank or position, pronounced

	"first", "second" etc.
out-of-band	Signaling which uses a different channel from the channel carrying voice.
parsed character data	XML text which is examined looking for tags and entities.
PBX	See Private Branch Exchange.
PCM	See Pulse Code Modulated.
Phone Markup Language	A markup language which was a precursor to VoiceXML.
phoneme	A phoneme is the "atom" of pronunciation, it is a single distinguishable sound. In English, a phoneme corresponds roughly to a consonant or vowel, that is, individual letters such as "d" and "a", or letter combinations like "ee" and "sh".
Plain Old Telephone Service	Typical consumer phone service.
play-off	False-positive detection of a touch-tone due to frequencies found in a prompt played by the computer and echoed back through analog equipment.
PML	See Phone Markup Language.
POTS	See Plain Old Telephone Service.
precedence	A property of an operator in a programming language or grammar. It specifies the relative "strength" of an operator. Stronger operators are evaluated before weaker operators, overriding the default left-to-right evaluation order for an expression. For example, multiplication (*) is usually stronger than addition (+), so a+b*c is evaluated as a+(b*c).
premium service	A telephone service for which the customer is billed additional changes on top of the usual local and long-distance tools.
presentation layer	The part of an application which interacts with the user.

PRI ISDN	See Primary Rate Interface.
Primary Rate Interface	An ISDN service with 23 voice channels and one signaling channel.
Private Branch Exchange	Term for private telephone switch operated by a subscriber.
process phase	Step in the FIA which executes content associated with a dialog item. Typically this results in a prompt being played to solicit input from the user.
programmable switch	A switch which supports a CTI link.
prosody	The technical term for cadence.
PSTN	See Public Switched Telephone Network.
Public Switched Telephone Network	The global telephone network which can connect two phones anywhere in the world.
pull	A protocol or service driven by requests by a client to a server.
pulse dialing	Used to describe rotary dialing.
Pulse Code Modulation	Digitized audio with a fixed number of bits per sample where the numerical value of the sample represents the amplitude of the sound.
push	A protocol or service driven by a server to a client.
R1	Another name for MF signaling.
R4	Dialogic's direct-to-board API.
raw file	Another name for a headerless file.
Remote Procedure Call	A protocol where the name of a function and a list of arguments is passed from a client to a server computer for execution.
rendering	The process of taking a presentation layer description such as HTML and VoiceXML and presenting it to the user, this may involve for example displaying text and graphics on a screen or playing audio.

RIFF	The standard Windows format for multi-media files including Wave sound files.
root document	A second VoiceXML page which is loaded at the same time as the page currently being interpreted.
rotary phone	An old-style phone which has a circular dial instead of a touch-tone keypad.
RPC	See Remote Procedure Call.
RS-232 serial line	A type of cable used to connect two pieces of equipment, the most common use for RS-232 is to hook up a modem.
S.100	A C language API for computer telephony developed by the ECTF.
sample	A numerical value representing the amplitude of sound (air pressure) at one instance in time.
sampling rate	Number of samples per second.
schema	A specification of XML tags and attributes together with a syntax for how they may be arranged in a document. One way of specifying a schema is to use a DTD. VoiceXML defines a schema.
scope	A range within the source code of the application where a variable, event handler or other language element can legally be used.
score	A numerical value which indicates the relative confidence that a match has been found by a voice recognizer, typically expressed as an integer percentage 0 .. 100% or a floating-point value between 0.0 and 1.0.
segmentation	The process of identifying syllable, word, sentence and phrase boundaries in recognized speech.
seize	The technical term for the process of requesting a line from a switch in order to make a call. On a POTS line, you seize by going off-hook.
select phase	The step in the FIA where the browser determines the next field to visit.

semantic markup	Markup which indicates the meaning of text, for example it can be used to indicate which fields in a dialog should be set to which values when a particular part of a voice grammar is matched.
semantics	The meaning of a text.
Session Initiation Protocol	An IP telephony protocol, an alternative to H.323.
SGML	Standard Generalized Markup Language. A precursor to XML.
shadow variable	A variable associated with a VoiceXML form item which is created automatically by the browser. It has the same name as the item with a dollar sign ($). Properties of this variable contain useful information about the item.
Signaling System 7	A signaling protocol used internally by the telephone network.
single tone	A tone which consists of a single pure frequency, i.e. is a sine wave.
SIP	See Session Initiation Protocol.
SIVR	See Speaker Independent Voice Recognition.
smart switch	Another name for a programmable switch, i.e. a switch which supports a CTI link.
speaker identification	A voice recognition technology which identifies a particular user from the unique characteristics of his or her voice. Not currently supported by VoiceXML.
Speaker Independent Voice Recognition	Speech recognition technology that does not require prior knowledge of or training by the user.
speaker verification	Speaker identification, emphasizing reduction or elimination of false-negatives (so you are sure that if the system identifies a user, it really is that user).
spectral analysis	Analysis done by voice recognizers to determine the frequency components of the voice and how they vary with time.
speech grammar	A template which specifies a set of valid utterances.

Speech Objects	Proprietary term used by Nuance for Dialog Components.
speech recognition	Technology for converting voice to text.
Speech Recognition Grammar Format	An XML language for specifying speech grammars.
speech synthesis	Technology for converting text into audio.
Speech Synthesis Markup Language	An XML language for marking up text for speech synthesis to request characteristics of the voice such as speed, gender and so on.
SpeechML	A precursor of VoiceXML.
Speech-to-Text	Alternative term for speech recognition.
SRGF	See Speech Recognition Grammar Format.
SS7	See Signaling System Seven.
SSML	See Speech Synthesis Markup Language.
stateless protocol	A protocol where the server does not save any information about the history of a client session. HTTP is such a protocol, by design all HTTP commands are completely self-contained and independent of previous commands.
streaming audio	Audio provided in real-time as opposed to being pre-recorded. A typical example would be live commentary from a sporting event.
store and forward	See message store and forward.
subdialog	A VoiceXML dialog which is called like a subroutine.
supervised transfer	An automated call transfer where call progress analysis is used to attempt to determine whether the second call was completed before finishing the transfer.
switch	Telephony equipment capable of connecting two calls or two call legs.

T1

A digital trunk which carries 24 channels. Comes in two main types: robbed-bit, where all 24 are audio channels, and PRI ISDN, where there are 23 voice channel and one signaling channel. Also called DS1.

T3

Another term for DS3.

talk path

The chain of links carrying audio between the end-points of a call.

TalkML

A precursor to VoiceXML.

talk-off

A false-positive detection of a touch-tone due to frequencies in the caller's voice.

tapered prompting

A feature of automated voice systems where the prompt played varies the second, third.. time the caller reaches a given point in the application.

TAPI

See Telephony Applications Programming Interface.

TDM

See Time Division Multiplexing.

Telephony Applications Programming Interface

Microsoft Windows API for CT.

text normalization

The process used by speech synthesizers to resolve ambiguous text, i.e. text which may be pronounced different ways in different contexts, such as "Dr.".

Text-to-Speech

Alternative name for speech synthesis.

three-tier architecture

Splitting an application into logically separate layers for presentation (user input / output), business logic (the core algorithms of the application) and database access / transaction processing.

three-way calling

A service available to consumers to create three-party conference calls.

Time Division Multiplexing

A technique to carry multiple channels in a single bit stream by increasing the bit rate or, equivalently, by increasing the sample size.

time-slot

One channel carried by a TDM bitstream.

token

Smallest unit used to build a grammar.

toll-free	Call which is billed to the called party, not the originating caller.
touch-tone	Common term for a DTMF digit.
trunk	Generic term for a line provided by the phone company to carry one or more conversations.
TTS	See Text-to-Speech.
twisted-pair	The type of cable often used for phone lines. Twisting reduces radio and electrical interference.
Unicode	A 16-bit character set which covers most of the world's major written languages and has been widely adopted by operating system and application vendors.
universal grammars	Grammars defined by the VoiceXML browser.
unsupervised transfer	Another term used for blind transfer.
URI	Uniform Resource Identifier. A generalized URL.
URL	Uniform Resource Locator. An Internet naming scheme for files and other resources.
URL encoding	A method of storing variables names and values into a text string. A typical string looks like $a=1\&b=2\&c=3$. Characters which are not legal in a URL are represented by $\%xx$ where xx is the hex value of the ASCII code. The string may be appended to a URL following a question mark "?" or stored in the Entity Body of an HTTP POST command.
user-directed	A conversation where the user asks questions or issues commands and the computer responds to them. A typical example is a pull-down menu in a GUI or a voice command such as "open accounts file".
user-initiative	Another term for user-directed.
valid document	Technical term for a XML document which matches a given schema.
value (of a tag)	In XML, the value of a tag is the text and/or subtags, if any, which appear between matching

start and end tags. Also referred to as the content of a tag.

VAP See Vox Array of Prompts.

VBase40 Another term for an IPF file.

visit A technical term used in the FIA to describe when the VoiceXML browser starts interpreting a given item in a dialog.

vocabulary Set of words or phrases known to a person or machine. Sometimes used specifically to a set of pre-recorded sounds which are used to build phrases which speak variable values, such as "you have" "twenty" "new messages".

voice browser Software which interacts with a user via voice input and output, driven by instructions from a Web server.

voice bus An internal bus in a CT system which carries real-time voice between different components.

voice modem A low-end, single-channel device which adds interactive voice capabilities to a data modem.

Voice over IP Technology which carries voice over an IP network.

voice recognition Technology which converts speech into text.

voice stop Voice recognition technology which allows a user to interrupt a prompt. Is sometimes used as a synonym for barge-in, sometimes it means a restricted capability where the user's voice stops a prompt, but that part of the speech is not available for recognition.

VoiceXML Forum An industry group which defined VoiceXML 1.0 and continues to evangelize the language.

VoIP See Voice over IP.

Vox Name for a de-facto standard set of file formats defined by Dialogic. There are six different file types; all are headerless so other formats are generally preferred where supported.

Vox Array of Prompts	Another term for an IPF file.
VoxML	A precursor of VoiceXML.
VR	Voice Recognition.
W3C	See World Wide Web Consortium.
Wireless Application Protocol	A protocol for sending Web pages to a small text display on a telephone. The pages are formatted using WML. WAP is to HTTP as WML is to HTML. Unlike VoiceXML, requires a special telephone.
Wireless Markup Language	XML language for WAP phones. See Wireless Application Protocol.
WML	See Wireless Markup Language.
WAP	See Wireless Application Protocol.
Wave	The Windows standard sound file format.
well-formed document	An XML document which conforms to the general XML syntax rules, it may or may not conform to a particular schema.
World Wide Web Consortium	A standards body for the Web.
Worldbet	A standard phonetic alphabet
XML Stylesheet Language	An XML language designed for transforming XML documents between different schemas. Also supports plain text, HTML and other non-XML formats.
X-SAMPA	A standard phonetic alphabet.
XSL	See XML Stylesheet Language.

Index